Intelligent Systems Reference Library

Volume 280

The aim of this series is to publish a Reference Library, including novel advances and developments in all aspects of Intelligent Systems in an easily accessible and well structured form. The series includes reference works, handbooks, compendia, textbooks, well-structured monographs, dictionaries, and encyclopedias. It contains well integrated knowledge and current information in the field of Intelligent Systems. The series covers the theory, applications, and design methods of Intelligent Systems. Virtually all disciplines such as engineering, computer science, avionics, business, e-commerce, environment, healthcare, physics and life science are included. The list of topics spans all the areas of modern intelligent systems such as: Ambient intelligence, Computational intelligence, Social intelligence, Computational neuroscience, Artificial life, Virtual society, Cognitive systems, DNA and immunity-based systems, e-Learning and teaching, Human-centred computing and Machine ethics, Intelligent control, Intelligent data analysis, Knowledge-based paradigms, Knowledge management, Intelligent agents, Intelligent decision making, Intelligent network security, Interactive entertainment, Learning paradigms, Recommender systems, Robotics and Mechatronics including human-machine teaming, Self-organizing and adaptive systems, Soft computing including Neural systems, Fuzzy systems, Evolutionary computing and the Fusion of these paradigms, Perception and Vision, Web intelligence and Multimedia.

Indexed by SCOPUS, DBLP, zbMATH, SCImago.

All books published in the series are submitted for consideration in Web of Science.

Jair Minoro Abe

An Introduction to Curry Systems

A Way to Algebraize Paraconsistent, Paracomplete, and Non-Alethic Logic

Springer

Jair Minoro Abe
Paulista University
São Paulo, Brazil

ISSN 1868-4394 ISSN 1868-4408 (electronic)
Intelligent Systems Reference Library
ISBN 978-3-031-97809-8 ISBN 978-3-031-97810-4 (eBook)
https://doi.org/10.1007/978-3-031-97810-4

This Springer imprint is published by the registered company Springer Nature Switzerland AG
The registered company address is: Gewerbestrasse 11, 6330 Cham, Switzerland

In memory of Miyuki (1958–2024)
To my beloved sister, now embraced by eternal light.
Your soul lives on in peace, and your love echoes in every breath of grace.

Preface

The present monograph deals with an algebraic version of the paraconsistent systems studied by da Costa, the well-known C_n, C_n* systems, and the paracomplete P_n, P_n* systems, non-alethic systems N_n, N_n*, and Pτ logics. The first manuscript (mimeographed) that the author is aware of, the algebraic version of the C_n paraconsistent systems, is the monograph of Costa, Álgebras de Curry, IME-USP, dated 1966. In the same monograph, the first articles on the subject, submitted in the Les Comptes Rendus de l'Académie des Sciences, on non-monotonic operators in lattices is published, and others. Later, around the 90s of the last century, Prof. Newton Costa, my doctoral advisor at the time, proposed to me an essay on orderly systems (three volumes), two of them by Prof. Constantino M. de Barros and the other, Curry Systems, where we would recover the theme, with details and additions by the present author. Since then, I have made some extensions to the calculus of system predicates over the years to obtain algebraic versions of the referred systems. Such results were based on P. Halmos' elegant monadic algebras.

P. Rosembloom presents in his book an algebraic treatment of classical propositional logic related to our studies. Its treatment is somewhat unorthodox; however, his work is referred to by Halmos.

We are aware that there are other ways of algebraizing paraconsistent systems, e.g., following Blok and Pigozzi and others. The techniques discussed here do not apply to specific non-classical systems but also to the classic logic itself, as shown in a solo work by M. Eytan, which we have included in this volume.

Thus, this book collects works from two generations: those of Prof. Newton Costa and those of the present author.

The author wishes to make it clear that this volume is only a first version of the subject, being an introductory text, and that improvements will be made if the work deserves future editions.

São Paulo, Brazil — Jair Minoro Abe

Contents

Chapter 1
Curry Systems

1.1 Introduction

This chapter presents the importance of studying algebraic versions of non-classical logical systems, especially paraconsistent, paracomplete, and non-alethic logics. The observations are general and do not go into too much depth due to the nature of the book. For historical aspects of paraconsistent logics, see [1].

1.1.1 The Importance of Algebraization in Non-classical Logics

The algebraization of logical systems has played a fundamental role in developing both classical and non-classical logics, offering a formal and structural framework through which the underlying principles of these systems can be analyzed and understood. While classical logic enjoys a long history of algebraic treatment through structures such as Boolean algebras, the algebraization of non-classical logics, including modal, intuitionistic, and paraconsistent logics, has been equally transformative. By providing a formal bridge between syntax and semantics, algebraization allows for a deeper exploration of logical systems, offering new tools for their analysis and application in a variety of domains.

Non-classical logics diverge from classical logic by modifying or rejecting certain foundational principles such as the Principle of Excluded Middle, or the Principle of Non-contradiction. This deviation gives rise to a more flexible reasoning framework capable of handling vagueness, uncertainty, contradictions, and other phenomena that classical logic cannot adequately address. These logics find applications in fields such as computer science, artificial intelligence, linguistics, and philosophy. However, as

J. M. Abe, *An Introduction to Curry Systems*, Intelligent Systems Reference Library 280, https://doi.org/10.1007/978-3-031-97810-4_1

non-classical logics is becoming prominent, algebraic methods have become indispensable for their systematic study and development. A reference for classical logic, see [2, 3].

Algebraization provides several key advantages in the study of non-classical logics. First, it facilitates the formalization of logical systems through well-defined mathematical structures, which allows for precise reasoning about their properties. Such observation is evident in how specific algebraic frameworks have successfully captured various non-classical logics. For example, intuitionistic logic is naturally captured by Heyting algebras, which generalize Boolean algebras by relaxing the Principle of Excluded Middle; see [2, 4–6]. Similarly, modal logics have been algebraized using modal algebras or Kripke frames representing the accessibility relations between possible worlds; see Goldblatt [7], Jónsson and Tarski [8].

Second, algebraic structures provide tools for exploring the metatheoretical properties of logical systems. Through algebraization, important properties such as soundness, completeness, and decidability can be studied in purely algebraic terms. For instance, the algebraic semantics of modal logics allow researchers to prove completeness theorems, showing that every valid formula has an algebraic counterpart in its associated frame; see Chagrov and Zakharyaschev [9]. The interplay between algebraic and relational semantics in modal logic further demonstrates the versatility of algebraic methods.

Another important benefit of algebraization is that it enables the development of new logical systems by modifying existing algebraic frameworks. In the case of paraconsistent logics, for example, algebraic structures such as de Morgan algebras and lattices with involution have been instrumental in formalizing systems that tolerate contradictions without collapsing into triviality; see Priest [10]. These structures allow for the coexistence of contradictory elements while preserving non-trivial inferential rules, a feat that is impossible within classical frameworks. Algebraic tools have also led to the classification of paraconsistent systems based on their algebraic properties, enabling a more nuanced understanding of their behavior.

Furthermore, algebraization plays a crucial role in developing computational models for non-classical logics. Algebraic semantics provide a rigorous foundation for developing decision procedures and automated theorem provers in fields like artificial intelligence and automated reasoning, where logics are often implemented in algorithms. For example, using algebraic models in fuzzy logic has enabled the creation of efficient algorithms for approximate reasoning in systems that deal with vague or imprecise information; see Hájek [11]. Translating logical systems into algebraic terms has thus proven essential in creating practical tools for reasoning in real-world applications.

Algebraization also aids in the study of duality theorems, which reveal deep connections between algebraic and relational semantics. For instance, the well-known duality between Boolean algebras and Stone spaces is a cornerstone of classical logic's algebraic treatment; see Johnstone [12]. Similar dualities have been explored in non-classical logics, such as the duality between Heyting algebras and specific topological spaces in intuitionistic logic; see Goldblatt [7]. These dualities enhance our understanding of logical systems and facilitate the transfer of techniques and results between algebra and logic.

In conclusion, the algebraization of non-classical logics provides a robust framework for analyzing, classifying, and applying these systems across various domains. By formalizing logical systems through algebraic structures, researchers can rigorously investigate their properties, establish connections between different systems, and develop new logics tailored to specific applications. The interplay between algebra and non-classical logics continues to be a rich and productive area of research, opening new avenues for theoretical exploration and practical implementation.

1.1.2 Introduction to the Algebraization of Paraconsistent Logics

The study of paraconsistent logics arises from the need to address inconsistencies within formal systems without falling into triviality. Strictly speaking, a theory based on a given logic is considered trivial if all of its (closed) formulas are theorems. Otherwise, the theory is said to be non-trivial. In traditional logic, where the Principle of Non-Contradiction (or Principle of Contradiction) applies, if it admits a contradiction (i.e., if both a proposition P and its negation ¬P are true), then any proposition Q can be derived as true in this system. Mathematically, this can be expressed as:

$$P \wedge \neg P \vdash Q$$

If a system allows P and ¬P to be simultaneously true, then any proposition Q can be derived (*ex contradictione sequitur quodlibet*).

Paraconsistent systems were developed to avoid this principle. They are novel logical systems that allow contradictions to exist without the entire system becoming trivial.

As noted above, algebraization is a crucial aspect of logical systems, bridging syntactic formalism and semantic structure. By transforming logical systems into algebraic structures, researchers can leverage powerful mathematical tools to analyze properties, behaviors, and relationships within these systems. In the case of paraconsistent logics, the process of algebraization not only facilitates more profound insights into their theoretical foundations but also allows the rigorous exploration of their practical applications.

The algebraization of paraconsistent logic involves mapping logical operations to corresponding algebraic entities, where contradictions can be effectively handled within algebraic structures. This process often relies on non-classical algebraic structures, such as de Morgan algebras, Heyting algebras, lattices, and semi-lattices, which provide the flexibility needed to capture the unique behavior of paraconsistent reasoning. These algebraic structures allow for a granular interpretation of truth values, supporting systems where a proposition and its negation can be true without leading to logical collapse [2, 4–6, 8, 13, 14].

This formal algebraic approach also supports investigating the metalogical properties of paraconsistent logics, such as completeness, soundness, consistency, and decidability; see Goldblatt [15]. It enables the development of automated reasoning tools and decision procedures, further advancing the applicability of paraconsistent systems in real-world scenarios, including information systems, legal reasoning, and inconsistent data handling.

This work embarks on a comprehensive study of the algebraization of paraconsistent logics via Curry Systems, systematically exploring various algebraic structures that accommodate the peculiarities of these non-classical logics. By delving into the structural properties of these algebraic systems, this book aims to contribute to the theoretical and applied aspects of paraconsistent reasoning, offering new insights into how contradictions can be managed within a mathematically sound framework.

In doing so, we hope to highlight the elegance of paraconsistent logics and their potential as powerful tools for reasoning in complex and inconsistent and/or paracomplete environments.

This chapter generalizes some basic concepts for studying Curry systems: pre-algebras, Curry algebras, pre-ordered systems, pre-lattices, Boolean pre-algebras, and pre-filters. They form the basis for subsequent chapters.

1.1.3 Introducing Basic Ideas for the Algebraization of Non-classical Logics

In the field of logic, mathematical structures play an essential and foundational role, particularly in the subdomain of algebraic logic. This role is supportive and often constitutive, as these structures provide the formal scaffolding for developing, understanding, and applying logical systems. Such observation is especially evident in the study of classical logic, where algebraic methods have been used to systematize and algebraize logical reasoning. Specifically, structures such as semi-lattices, Boolean algebras, monadic algebras, and polyadic algebras, alongside topological spaces, have been employed extensively in classical algebraic logic. For basic algebraic structures seen in this theme, see Grätzer [14], Henkin [16], Iséki [2, 4], Halmos [5], Henkin, Monk and Tarki [6], and Sikorski [17].

The typical route to creating algebraic structures in classical logic is through a process called passage to the quotient. In this approach, given a logical system, one selects a suitable equivalence relation compatible with the fundamental logical operations. Once such a relation is identified, the resulting quotient structure algebraizes the original logical system. This process demonstrates that Boolean algebras and Heyting algebras correspond to the algebraizations of classical propositional calculus and intuitionistic propositional calculus, respectively. In these cases, the equivalence relation ensures that logical operations are preserved in the algebraic structure, making these algebras highly useful for studying logical systems.

However, while effective in classical logic, this approach often encounters limitations in non-classical logics. Identifying a significant congruence relation is not always possible in some instances, such as in many paraconsistent logical systems. A congruence relation, by definition, is an equivalence relation that respects the operations of the logical system; yet, in some systems, no such relation exists that is useful and aligned with the underlying logic. Such property poses a fundamental challenge to the traditional methods of algebraization, as there is no straightforward way to apply the quotient procedure in the absence of a congruence. Paraconsistent logic, which allows for the coexistence of contradictory statements without collapsing into triviality, exemplifies such a situation. In these systems, classical methods of algebraization are inadequate, and new approaches are required.

Even within classical logic, where congruence relations are more readily available, passing to the quotient is not always the most appropriate or desirable method. In some cases, the process of quotienting obscures critical logical insights. For example, in classical propositional calculus, concepts such as Smullyan's tableaux and Hintikka's sets offer alternative perspectives on logical reasoning that are not easily captured through algebraic structures derived from the quotient process. Smullyan's tableaux provide a syntactic method for determining the satisfiability of logical formulas, while Hintikka's sets are used in semantic tableaux and model-theoretic approaches. These techniques rely on specific structural properties of the logical system, which can be masked or lost in the transition to a quotient structure.

This recognition of the limitations of traditional algebraic methods has led to the development and use of pre-algebras, which provide a more flexible framework for dealing with logical systems. Pre-algebras are structures that retain certain algebraic properties without fully satisfying the conditions of a quotient algebra. The use of pre-algebras is particularly relevant in two key scenarios.

(1) When no reasonable congruence relation exists (or is known), as may occur when the chosen equivalence relation is incompatible with all logical operations. In these cases, traditional algebraic techniques fail to capture the essential features of the logical system.
(2) When a congruence relation is available, but passing to the quotient is undesirable due to the risk of obscuring significant logical facts. Here, the quotient may simplify the structure too much, losing critical information necessary for a deeper understanding of the system.

In the first scenario, where no congruence relation exists, Curry algebras come into play. Named after the American logician H.B. Curry, these algebras offer a way to handle logical systems that defy traditional algebraic methods. Curry algebras retain enough structure to allow for meaningful algebraic manipulation, even without congruences, making them particularly useful for non-classical logics like paraconsistent systems. In the second scenario, where congruence relations are present but undesirable, pre-algebras provide an alternative to full quotient structures, allowing one to work within a framework that preserves more of the logical system's original complexity; see Curry [18–20].

The introduction of the Curry system serves to formalize a general theory of the algebraization of logical systems. The Curry system encompasses all the structures of interest in the mathematical treatment of logic, offering a unifying framework that is flexible enough to account for both classical and non-classical systems. By modifying or enriching the concept of a Curry system, one can derive specific structures such as logical matrices, Kripke models, and the models used in classical model theory. While not always directly related to algebraization problems, these structures fall within the broader scope of the Curry system and highlight its versatility.

The importance of the Curry system concept is underscored by its ability to suggest new structures for logical analysis. For example, one might consider structures that feature multiple equivalence relations, one of which could be equality. These structures are closely related to the theory of logical matrices and multi-valued logics, which have applications in fields as diverse as computer science, linguistics, and philosophy. Multi-valued logics, for instance, generalize classical two-valued logic by allowing for more than two truth values, and their algebraic treatment requires new kinds of structural tools that the Curry system framework is well-equipped to handle.

For these reasons, the structures discussed here have been named after H.B. Curry, not only to honor his contributions to the field of logic but also because he was a key proponent of the systematic use of pre-algebraic structures. Curry's work laid the foundation for the modern understanding of how algebraic structures can be used to model logical systems in a way that transcends the limitations of classical methods. His insights into using pre-algebraic techniques have proved invaluable, especially in studying non-classical logics.

In conclusion, without much exaggeration, it can be argued that the study of logic from a mathematical perspective is mainly reducible to the study of Curry systems. These systems offer a comprehensive framework that integrates both classical and non-classical logics, providing the tools necessary for a deeper and more nuanced understanding of logical systems. As logic evolves, the Curry system concept will likely play a more central role in the algebraic study of logic, driving discoveries and insights.

1.2 Curry Systems

In the sequence, we present the main concepts of the Curry system.

Definition 2.1 Let A be a non-empty set. It is said that f is an n-ary operation in A if each n-tuple $(x_1, x_2,\ldots, x_n)$ of elements of A, f matches a single element x of A. When this occurs, we write.

$$f(x_1, x_2, \ldots, x_n) \equiv x$$

If n = 1 or n = 2, f is said, respectively, an operator (or unary operation) or a binary operation. In the latter case, we also write $x_1 f x_2 \equiv x$ instead of the previous notation.

Definition 2.2 Suppose that the equivalence relation relação is defined in A. The n-ary operation f is said to be i-compatible with ≡, $1 \leq i \leq n$, if any $x_1, x_2, \ldots, x_{i-1}, x_{i+1}, \ldots, x_n$ belonging to A, $a \equiv b$ entails

$$f(x_1, x_2, \ldots, x_{i-1}, a, x_{i+1}, \ldots, x_n) \equiv f(x_1, x_2, \ldots, x_{i-1}, b, x_{i+1}, \ldots, x_n)$$

Operation f is said to be compatible with ≡ (or monotonous concerning ≡), if, for all i, $1 \leq i \leq n$, j is i-compatible with ≡.

Definition 2.3 Suppose that in A, the equivalence relation ≡ is defined. Let R be an n-ary relation over A. It is said that R is compatible with ≡ (or monotonous concerning ≡) if, for any $x_1, x_2, \ldots, x_n, x_1', x_2', \ldots, x_n'$ belonging to A, if $(x_1, x_2, \ldots, x_n) \in R$, $x_i \equiv x_i'$, $1 \leq i \leq n$, then $\left(x_1', x_2', \ldots, x_n'\right) \in R$.

Definition 2.4 The n-ary operation f in A is called ≡-idempotent if, for all $x \in A$, we have $f(x, x, \ldots, x) \equiv x$. If f is ≡-idempotent, f will be said to be idempotent.

Definition 2.5 By a Curry system, we mean a structure $\left[A, (\equiv)_i, (S)_j, (R)_k, (j)_l, (C)_m\right]$, where,

(A_1) $A \neq \emptyset$
(A_2) $(\equiv)_i$ is a collection of equivalence relations over A.
(A_3) $(S)_j$ is a family of subsets of A.
(A_4) $(R)_k$ is a finite relations collection defined in A.
(A_5) $(f)_l$ is a family of operations defined in A.
(A_6) $(C)_m$ is a finite collection of elements of A.

In general, the operations, relations, etc., of a Curry system satisfy certain conditions of algebraic "nature." For example, operations can be commutative, associative, etc.

In what follows, we only consider Curry systems in which the relation $(\equiv)_i$ contains a single equivalence relation ≡, which we will call the fundamental (equivalence) relation.

Definition 2.6 A Curry system is said to be a pre-algebra if the collection $(S)_j$ of subsets of A is empty and if all the operations and relations of that system are monotonous concerning ≡. In this case, ≡ is a congruence relation over A. Usually, in pre-algebras, the family $(R_k)_{k \in K}$ of relations is also empty. In the particular case where the fundamental relation ≡ is equality, pre-algebra is algebra in the usual sense.

Definition 2.7 A Curry system is called a Curry Algebra if it has at least one non-monotonic operation or relation.

Definition 2.8 A Curry system is called a pre-ordered Curry system if the collection $(S_j)_{j\in J}$ of subsets A is empty, and one of the relations $\leq$ of the collection $(R_k)_{k\in K}$ is pre-order on A, i.e., the following conditions apply.

(R_1) For all $x \in A$, there is $x \leq x$ (reflexivity).
(R_2) For all $x, y, z \in A$, if $x \leq y$ and $y \leq z$, then $x \leq z$ (transitivity).
(R_3) For all $x, y, x', y' \in A$, if $x \leq y$, $x' \equiv x$ and $y' \equiv y$, then $x' \equiv y'$.

Definition 2.9 Let $\equiv$ and $\equiv_1$ equivalence relations over a set A. It is said that the relation $\equiv_1$ is finer than the relation $\equiv$ if $\equiv_1$ contains $\equiv$, which is equivalent to saying that, if $x, y \leq A$, if $x \leq y$, then $x \equiv_1 y$.

In a pre-ordered Curry system, we define the relation $\equiv_1$ as follows: $x \equiv_1 y$ and only if $x \leq y$ and $y \leq x$, for any $x, y \in A$. It is easily shown that $\equiv_1$ is an equivalence relation, which we call the equivalence relation associated with the pre-order $\leq$.

Lemma 2.10 With the above notations, in a pre-ordered Curry system A, the relation $\equiv_1$ associated with the pre-order $\leq$ is finer than the fundamental relation $\equiv$ of that system.

Proof Indeed, let $x, y \in A$ be such that $x \equiv y$. For the reflexivity of $\leq$, we have $x \leq x$. Also, $x \leq x$ (since $\equiv$ is an equivalence relation). Therefore, by R_2, it follows that $x \leq y$. Similarly, $y \leq x$ and therefore $x \equiv_1 y$.

Observation 2.11 It should be noted that, in general, in a pre-ordered Curry system, the relation $\equiv_1$ associated with the pre-order $\leq$ is strictly finer (i.e., $\equiv \subseteq \equiv_1$) than the fundamental relation $\equiv$.

Definition 2.12 A pre-ordered Curry system is called a pre-ordered system if the equivalence relation $\equiv_1$ associated with the pre-order $\leq$ coincides with the fundamental relation $\equiv$.

The above definition assures us that we can suppress the fundamental relation $\equiv$ of the system, obtaining a pre-ordered system in the usual sense (i.e., a pair [R, $\leq$] such that $\leq$ is reflective and transitive).

1.3 Pre-ordered Systems

Let [R, $\leq$] be a pre-ordered system. If x and y are any elements of R, and $x \leq y$, we say "x is less than y" or "x precedes y". If $x \leq y$, we also write $y \geq x$, and it says "y is greater than x" or "y succeeds x."

Definition 3.1 Let [R, $\leq$] be a pre-ordered system. If A is a subset of R, an upper bound for A in R (or majorant for A in R) is an element a of R such that $x \leq a$, for all $x \in A$. Similarly, $b \in$ R is a lower bound (or minorant) for A in R if for all $x \in A$, $b \leq x$.

Definition 3.2 A pre-ordered system [R, $\leq$] is called a totally pre-ordered system if, for any x, $y \in$ R, or $x \leq y$ or $y \leq x$.

Definition 3.3 Let [R, $\leq$] be a pre-ordered Curry system and A a subset of R. $x \in$ R is said to be a supremum (or lower upper bound) of A if x is an upper bound for A in R and is a lower limit for the set of upper limits for A in R. If analogously, $x \in$ R is called the infimum of A if x is a lower limit for A in R and greater than or equal to any other lower limit for A in R.

Observation 3.4 Let [R, $\leq$] be a pre-ordered system of Curry and $A \subseteq A$. The supremum of A (if any) may not be unique. Let x and y be supremum for A. Since x is supremum for A, $x \leq y$. Similarly, we have $y \leq x$. Due to generalized antisymmetry, it follows that $x \equiv y$, which may not mean identity.

The previous observation shows us that a given subset's supremum (if any) may not be unique, and this property differs from ordinary ordered systems.

We denote the set of the supremum of A by Sup(A) and the set of the infimum of A by Inf(A). If $A = \{a_i | \, i \in Z\}$ is an indexed collection of elements of R, we write $\vee_{i \in I} a_i$ for a fixed element of Sup (A) and $\wedge_{i \in I} a_i$ for a fixed element of Inf (A). If A is the set consisting of two elements $\{x, y\}$, we write $x \vee y$ for a fixed element of Sup(A) and $x \wedge y$ for a fixed element of Inf(A).

Theorem 3.5 Let [R, $\leq$] be a pre-ordered Curry system. Let us consider the quotient set R /$\equiv$ (where $\equiv$ is the relation associated with $\leq$) and define the relation $\leq$ "over R/$\equiv$ as follows:

[a] $\leq$"[b] if and only if $a \leq b$. We claim that the algebraic structure [R/$\equiv$, $\leq$, '] is an ordered system.

Proof Indeed, let us show that $\leq$ 'is well defined. Let a', b' $\in$ R, such that $a \equiv a$' and $b \equiv b$'. Then, $a \equiv b$ if and only if a'$\equiv$ b' (by R). Therefore, the definition of $\leq$ 'is independent of elected representatives a, and b. The reflexivity and transitivity of $\leq$ 'are immediate. Let a, $b \in$ R such that [a] $\leq$ '[b] and [b] $\leq$ '[a]. Therefore, $a \leq b$ and $b \leq a$, and, therefore, $a \equiv b$. However, this implies that [a] = [b], because $\equiv$ is an equivalence relation. Consequently, $\leq$ ' is antisymmetric.

1.4 Pre-lattice

Definition 4.1 A pre-lattice is a pre-ordered Curry system [R, $\leq$] satisfying the following condition.

(R4) For any x, $y \in$ R, sup($\{x, y\}$) $\neq \emptyset$ and inf($\{x, y\}$) $\neq \emptyset$.
Thus, in a pre-lattice R, for any x, $y \in$ R, there are $x \wedge y$ and $x \vee y$.

Lemma 4.2 Every totally pre-ordered Curry system [R, $\leq$] is a pre-lattice.

Proof For any elements $x, y \in \mathrm{R}$, $x \leq y$ or $y \leq x$. Suppose $x \leq y$. Since $\leq$ is reflexive, $x \leq x$. So, x is a lower limit of $\{x, y\}$. If $x' \in \mathrm{R}$ is another lower limit of $\{x, y\}$, it follows that $x' \leq x$, which means that $x \in \inf \{x, y\}$. In the same way, it is shown that $y \in \sup \{x, y\}$.

Lemma 4.3 Any non-empty finite subset of a pre-lattice admits a supremum and an infimum.

Proof Immediate.

Definition 4.4 Let m be an infinite cardinal. An m-complete pre-lattice is a pre-lattice in which each subset of cardinal elements less than or equal to m has a supremum and infimum. A pre-lattice is $< m$-complete if each set of cardinal elements strictly less than m has a supremum and an infimum. A pre-lattice is complete if it is m-complete for each infinite cardinal m.

Lemma 4.5 Each pre-lattice is $< \aleph_0$-complete.

Proof A consequence of the previous lemma.

Consider a pre-lattice $[\mathrm{R}, \leq]$. We define in R two operations $\wedge$ and $\vee$:

$$x \wedge y \equiv z, \text{ where } z \in \inf(\{x, y\})$$

$$x \wedge y \equiv w, \text{ where } w \in \sup(\{x, y\})$$

Theorem 4.6 With the above notations, operations $\wedge$ and $\vee$ have the following properties:

(L_1) $x \vee y \equiv y \vee x$
(L_2) $x \vee x \equiv x$
(L_3) $x \vee (y \vee z) \equiv (x \vee z) \vee z$
(L_4) $(x \wedge y) \vee y \equiv x$
(L_5) $x \wedge y \equiv x$
(L_6) $x \wedge (y \wedge z) \equiv (x \wedge z) \wedge z$
(L_7) $x \vee y \wedge y \equiv y$
(L_8) $x \wedge y \vee y \equiv y$
(L_9) (L_9)
If $x \equiv x'$ *and* $y \equiv y'$, then $x \wedge y \equiv x' \wedge y'$ and $x \vee y \equiv x' \vee y'$.

Proof (L1). Of course, $\sup(\{x, y\}) \equiv \sup(\{y, x\})$. Therefore, let $z \in \inf(\{x, y\})$ and $z' \in \inf(\{y, x\})$. So, $z \equiv z'$. In fact, $x \vee x = t$, $t \in \sup(\{x, x\})$. It is clear that $x \in \sup(\{x, x\})$, hence $t \equiv x$, that is, $x \vee x \equiv x$.

(L3) Let

1. $y \vee z = t$, $t \in \sup(\{y, z\})$,
2. $x \vee (y \vee z) = u$, $u \in \sup(\{x, y \vee z\})$,
3. $x \vee y = v$, $v \in \sup(\{x, y\})$,

4. $(x \vee y) \vee z = w,\ w \in \sup(\{x \vee y, z\})$.

We have:

5. $u \geq x$ 2
6. $u \geq y \vee z = t$ 2
7. $t \geq y$ 1
8. $t \geq z$ 1
9. $u \geq y$ 6, 7
10. $u \geq z$ 6, 8
11. $v \geq x$ 3
12. $v \geq y$ 3
13. $\omega \geq z$ 4
14. $w \geq v$ 4
15. $w \geq x$ 11, 14
16. $w \geq y$ 12, 14
17. $v \leq u$ 2, 6, 7, 3
18. $z \leq u$ 6, 8
19. $w \geq u$ 17, 18, 4
20. $t \geq w$ 16, 13, 1
21. $u \geq w$ 15, 20, 2
22. $w \geq u$ 19, 21

Thus, $x \vee (y \vee z) \equiv (x \vee y) \vee z$.

(L4) Similar to (L1)

(L5) Similar to (L2)

(L6) Similar to (L3)

(L7) Let it be

1. $x \vee y = u,\ u \in sup\{x, y\}$
2. $x, y \leq u$ 1
3. $y \leq u, y$ 2, R_1
4. if $t \leq u, y$, then, $t \leq y$
5. $y \in \inf\{x, y\}$

Thus,$(x \wedge y) \wedge y \equiv y$.

(L8) Similar to (L7)

(L9) $x \equiv x', y \equiv y', t \in \inf\{x, y\}, t' \in \inf\left\{x', y'\right\}$

1. $t \equiv x \wedge y$
2. $t' \equiv x' \wedge y'$
3. $t \leq t'$
4. $t \leq t', y'$ 1, 3, R_5
5. $t \leq t'$ 2, 4
6. Similarly $t' \leq t$

7. $t \equiv t'$, i.e., $x \wedge y \equiv x' \wedge y'$ 5, 6
8. Analogously we have $x \vee y \equiv x' \vee y'$.

Corollary 4.6.1 With the notations of the previous theorem, we have:
$(L_{10})x \equiv x \wedge y$ if and only if $y \equiv x \vee y$

Proof Suppose $x \equiv x \wedge y$. So,

1. $x, y \leq x$ hypothesis
2. $x \leq x$ 1 hypothesis
3. $y \leq y$ R_2
4. $x \vee y \leq y$ 2, 3
5. $y \leq x \vee y$ L7
6. $v \equiv x \vee y$

Conversely, suppose $y \equiv x \vee y$.

1. $x, y \leq y$ hypothesis
2. $x \leq x$ R_2
3. $x \leq x \leq y$ 1, 2
4. $x \wedge y \leq x$ L8
5. $x \equiv x \wedge y$ 3, 4, R3.

Theorem 4.7 Let [R, $\equiv \wedge, \vee$] be a Curry system such that $\equiv$ is the fundamental equivalence relation over R and $\wedge$, $\vee$ are two binary operations on A satisfying the conditions L2—L9 of the previous theorem. We define the relation $\leq$ in R by putting $x \leq y$ if $x \equiv x \wedge y$. So, the structure [R, $\leq$] is a pre-lattice.

Proof Let us check the pre-lattice axioms R_1—R_4.

(R_1) Indeed, $x \wedge x \equiv x$ per L5. Thus, $x \leq x$ for all $x \in$ R
(R_2) Let x, y, z $\in$ R be such that $x \leq y$ and $y \leq z$. So, $x \wedge x \equiv x$ and $y \equiv y \wedge$ z. Therefore $x \equiv x \wedge x = x \wedge (y \wedge z) \equiv (x \wedge y) \wedge z \equiv x \wedge z$ (by L9). Therefore, $x \leq z$.
(R_3) Let x, y, z $\in$ R be such that $x \leq y$, $x \equiv x$' and $y \equiv y$'. Therefore, $x' \equiv x$ *and* $x' \equiv x \wedge y$ (by (L9)). As, $x \wedge y \equiv x' \wedge y'$*and* $x' \equiv x' \wedge y'$*or* $x' \leq y'$
(R_4) Any two elements x, $y \in$ R, sup $\{x, y\} \neq \emptyset$ and inf $\{x, y\} \neq \emptyset$. Indeed, $x \equiv x \wedge (x \vee y)$ and $x \equiv y \wedge (y \vee x)$ (by L7). So, $x, y \leq x \vee y$. If z $\in$ R is such that $x, y \leq z$, that is $x \equiv x \wedge z$ e $y \equiv y \wedge z$, we have, for L10, that $z \equiv z \vee x$ and $z \equiv z \wedge y$. Consequently, $z \equiv z \vee z \equiv (z \vee x) \vee (z \vee y) \vee z \equiv (x \vee y)$ (by L2, L9 and L3). Again, applying L10, we obtain $x \vee y \leq z$, which results in $x \vee y \leq z$, that is, $x \vee y \equiv \sup\{x, y\}$.

We now prove that $x \wedge y$ is a infimum of $\{x, y\}$. We have to $x \equiv x \wedge x \wedge y \equiv (x \wedge x) \wedge y \equiv x \wedge (x \wedge y) \equiv (x \wedge y) \wedge x)$ and $x \wedge y \equiv x \wedge (y \wedge y) \equiv (x \wedge y) \wedge y$ by (L5, L6, L4). Therefore, $x \wedge y \leq x, y$. Let $z \in$ R be such that z $\leq x$, y, that is, $z \equiv z \wedge x$ *and* $z \equiv z \wedge y$. We have $z \equiv z \wedge z \equiv (z \wedge x) \wedge (z \wedge y) \equiv z \wedge (x \wedge y)$, hence $z \leq x \wedge y$ and therefore $x \wedge y \equiv \inf\{x, y\}$.

Corollary 4.7.1 Let [R, $\leq$] be a pre-lattice. Let [R, $\wedge$, $\vee$] be the structure obtained as the previous theorem indicates. Such structures are equivalent (in a sense that generalizes Bourbaki).

The relation between the [R, $\leq$] and [R, $\wedge$, $\vee$] structures can be outlined as follows:

$$[\mathrm{R}, \leq] \rightarrow_{\mathrm{Def}} [\mathrm{R}, \wedge, \vee] \rightarrow_{\mathrm{Def}} [\mathrm{R}, \leq']$$

$$[\mathrm{R}, \wedge, \vee] \rightarrow_{\mathrm{Def}} [\mathrm{R}, \leq] \rightarrow_{\mathrm{Def}} [\mathrm{R}, \wedge', \vee']$$

In the above scheme, we have: $[\mathrm{R}, \leq] = \left[\mathrm{R},' \leq'\right]$ and $[R, \wedge, \vee] = \left[R, \wedge', \vee'\right]$.

Proof Due to the previous theorem.

Because of the above considerations, from now on, we will represent a pre-lattice such as $[R, \leq]$ or $[R, \wedge, \vee]$.

Lemma 4.8 If X and Y are finite subsets of a pre-lattice [R, $\leq$], then Inf(X) $\wedge$ Inf(Y) = Inf(X $\cap$ Y).

Proof Immediate.

Theorem 4.9 Let [R, $\leq$] be a pre-lattice. Then, the ordered system[R/$\equiv$, $\leq$] obtained as indicated in Theorem 3.5 is a lattice in the usual sense.

Proof Indeed, be $[a]$, $[b] \in$ R/$\equiv$. We have to $\sup\{a, b\} \neq \emptyset$. Let $a \vee b \in \sup\{a, b\}$. We claim that $[a] \vee [b] = [a \vee b]$. Like a, b, $\leq a \vee b$, it follows that $[a]$, $[b] = [a \vee b]$. If $c \in$ R is such that $[a]$, $[b] = [c]$, it follows that a, $b \leq c$, hence $a \vee b \leq c$, that is, $[a \vee b] \leq [c]$. Therefore, $[a] \vee [b] = [a \vee b]$. Similarly, it is shown that $[a] \wedge [b] = [a \wedge b]$.

Note For language abuse, we are using the same symbol $\vee$ for both R and elements of R/$\equiv$. The same is true when identical symbols are used in different contexts, but the attentive reader will see no difficulty contextualizing them.

1.5 Complemented and Distributive Pre-lattices

We have already seen that if a pre-ordered Curry system [R, $\leq$] has upper limits, they are equivalent. In particular, this is true for any pre-lattice, and we call these elements maximum. We will denote one of the maximum elements that we fix by 1. Similarly, if a pre-lattice has lower bounds, they are equivalent. We will name these elements as minimum and denote any of them (fixed) by 0.

A maximum element of a pre-lattice must be distinguished from a maximal element. An element x of a pre-lattice is said to be a maximal element if no element of the pre-lattice is strictly greater than x.

Lemma 5.1 If a pre-lattice has maximum elements, then they are equivalent.

Proof Result of Lemma 3.4.

Definition 5.2 A pre-lattice $[R, \equiv, \leq]$ is called complemented if it has a maximum element 1, a minimum element 0, and for each $x \in R$, there is an element $y \in R$ such that:

(R5) $x \vee y \equiv 1$ and $x \wedge y \equiv 0$

Such an element y is called a complement of x. In a complemented pre-lattice, each element does not necessarily have equivalent complements.

We denote by $[R, \wedge, \vee, 0, 1]$ (or $[R, \leq, 0, 1]$) a pre-lattice with a maximum element 1 and a minimum element 0.

Lemma 5.3 Let $[R, \wedge, \vee]$ be a pre-lattice. We then have $(x \wedge y) \vee (x \wedge z) \leq x \wedge (y \vee z)$ and $x \vee (y \vee z) \leq (x \wedge y) \leq (x \vee z)$, for any $x, y, z \in R$.

Proof We have:

1. $s = y \vee z, s \in \sup\{y, z\}$.
2. $t = x \wedge \mathrm{s}, t \in \inf\{x, \mathrm{s}\}$.
3. $u = x \wedge y, u \in \inf\{x, y\}$.
4. $v = x \wedge z, v \in \inf\{x, z\}$.
5. $w = u \vee \mathrm{v}, w \in \sup\{y, z\}$.
6. $y, z \leq s \quad 1$
7. $z \leq x, y \quad 3$
8. $v \leq x, z \quad 4$
9. $\mathrm{u}, v \leq x \quad 6, 7, 8$
10. $t \leq x, s \quad 2$
11. $\mathrm{u}, v \leq x \quad 7, 8$
12. $\mathrm{u}, v \leq t \quad 9, 11, 2$
13. $w \leq t \quad 12, 5$

Therefore, $(x \wedge y) \vee (x \wedge z) \leq x \wedge (y \vee z)$. The proof of the other inequality is similar.

Definition 5.4 A pre-lattice $[R, \wedge, \vee]$ is called distributive if the following conditions are met:

(R6) $(x \vee y) \wedge z \equiv (x \wedge z) \vee (y \wedge z)$
(R7) $(x \wedge y) \vee z \equiv (x \vee z) \wedge (y \vee z)$

Theorem 5.5 A necessary and sufficient condition for a pre-lattice $[R, \equiv, \wedge, \vee]$ to be distributive is that:

(1) $x \wedge (y \vee z) \leq (x \wedge y) \vee (x \wedge z)$
(2) $(x \wedge y) \wedge (x \vee z) \leq \vee(y \wedge z)$

Proof It is an immediate consequence of the previous lemma.

Lemma 5.6 Each element has equivalent complements in a complemented and distributive pre-lattice.

Proof Let y and z be complementary to x. Then, $x \vee y \equiv 1$ and $x \wedge z \equiv 0$. We then have:

$$y \equiv y \vee 0 \equiv y \vee (x \wedge z) \equiv (y \vee x) \wedge (y \vee z) \equiv 1 \wedge (y \vee z) \equiv y \vee z$$

Similarly, we have $z \equiv z \vee y$ and, therefore, $y \equiv z$.

Observation 5.7 In any pre-lattice, condition R_6 is verified if and only if R_7 is verified.

In a complementary and distributive pre-lattice, we denote by x' a (fixed) complement of x.

1.6 Boolean Pre-algebras

Definition 6.1 A Boolean pre-algebra (or Boolean pre-algebra) is a complemented and distributive pre-lattice. In other words, a Boolean pre-algebra is a Curry system [B, ∧, ∨, ', 0, 1], satisfying the conditions R_1–R_7.

Lemma 6.2 Let 2 be the set formed by two elements 0 and 1. Let us define a partial order ≤ over 2 by 0' = 1 and 1' = 0. We also define an operator ' over 2 like 0' = 1 and 1' = 0. So, the system [2, 1, '] is a Boolean pre-algebra (which we will denote by 2).

Proof Immediate.

Theorem 6.3 Let [B, ∧, ∨ ', 0, 1] be a Boolean pre-algebra. Thus, the lattice B' = [B/≡, ∧, ∨ ', 0, 1] is a Boolean algebra in the usual sense.

Proof Let $a \in$ B. Then there is $a' \in$ B such that $a \vee a' \equiv 1$ and $a \wedge a' \equiv 0$. Therefore, $[a] \vee [a'] \equiv [9]$ and $[a] \wedge [a'] \equiv [0]$. Thus, B' is a complemented lattice, where $[a'] \equiv [a]'$.

We also have that for all $x, y, z, \in$ R

$$x \wedge (y \vee z) \equiv (x \wedge y) \vee (x \wedge z)$$

$$x \vee (y \wedge z) \equiv (x \vee y) \wedge (x \vee z)$$

So, it is easy to see that

$$[x] \wedge ([y] \vee [z]) \equiv ([x] \wedge [y]) \vee ([x] \wedge [z]) \text{ and}$$
$$[x] \vee ([y] \wedge [z]) \equiv ([x] \vee [y]) \wedge ([x] \vee [z]).$$

Therefore, B/≡ is a distributive and complemented lattice. That the above definitions are independent of representatives is immediate.

1.7 Pre-filters and Pre-ideals

Definition 7.1 Let [R, ≡, ≤] be a pre-lattice. A pre-filter in [R, ≡, ≤] is a non-empty subset F of R such that:

(1) For all $x, y \in F$, $x \wedge y \in F$
(2) For all $x \in F$ and $y \in R$, if $x \leq y$, then $y \in F$
(3) For all $x \in F$ and $y \in R$, if $x \equiv y$, then $y \in F$

Definition 7.2 Let [R, ≡, ≤] be a pre-lattice. A pre-ideal in [R, ≡, ≤] is a non-empty subset I of R such that:

(1) For all $x, y \in I$, $x \vee y \in I$
(2) For all $x \in I$ and $y \in R$, if $x \leq y$, then $y \in I$
(3) For all $x \in I$ and $y \in R$, if $x \equiv y$, then $y \in I$

Lemma 7.3 For each $x \in R$, the set $\{y \in R \mid x \leq y\}$ is a pre-filter, called the pre-filter generated by x. Similarly, $\{y \in R \mid y \leq x\}$ is a pre-ideal named the pre-ideal generated by x.

Proof Immediate.

Lemma 7.4 In any finite pre-lattice, every pre-filter and every pre-ideal is principal.

Proof Immediate.

Definition 7.5 A prefilter F in a pre-lattice [R, ≡, ≤] is called proper if F is a proper subset of R, that is, $F \neq R$. Of course, if R has a minimum of 0, F is proper if and only if $0 \notin F$. Similarly, a pre-ideal I is proper if $I \neq R$, and in a pre-lattice with, a maximum of 1, I is proper ideal if, and only if, $1 \notin I$.

Observation 7.6 Duality. In any Boolean pre-algebra, if under conditions R_1 - R_7 we change ∧ by ∨, ∨ by ∧, 0 by 1, and 1 by 0, identities are transformed into new true identities. Any "Boolean" proposition obtained from another Boolean proposition through the above transformation $y \leq x$ is called "dual" of the first. Therefore, the dual of any true proposition about Boolean pre-algebras "expressed" in terms of ∧, ∨, 0, 1, and ≡ is also a true proposition about Boolean pre-algebras.

Duality can be extended to involve the relations ≤ and ≥ .

Lemma 7.7 In a Boolean pre-algebra, $x \wedge x' \equiv 0$ if and only if $x \leq y$. Moreover, $x \vee x' \equiv 1$ if and only if $y \leq x$.

Proof Suppose $x \wedge y' \equiv 0$. Then,$x \equiv x \wedge 1 \equiv x \wedge (y \vee y') \equiv (x \wedge y) \vee (x \wedge y') \equiv (x \wedge y) \vee 0 \equiv x \wedge y \leq y$.

Conversely, if $x \leq y$, then $x \wedge y \equiv x$ and, therefore $x \wedge y' \equiv (x \wedge y) \wedge y' \equiv x \wedge (y \wedge y') \equiv x \wedge 0 \equiv 0$. The other part follows up on duality.

Definition 7.8 Let B be a Boolean pre-algebra. A subset F of B is said to have the *property of the finite intersection* (which we abbreviate by pfi) if, for any finite subset $\{a_1, a_2, \ldots, a_n\}$ of F, we have $\wedge^{n}{}_{i=1} a_i$ is distinct from 0.

Definition 7.9 Let B be a Boolean pre-algebra. Let $A \subseteq$ B. Consider the co-initial part generated by A, $R_+(A)$, i.e., $R_+(A) = \{x \in B \mid \text{for some } a \in B, a \leq x\}$.

Let us also consider the set $I(a)$ of the infimum of all finite subsets of A, that is, $I(a) \equiv \{x \mid x \in \text{Inf}(X) \text{ and } X \in \chi(A)\}$.

A base for a pre-filter F is a set A such that $R_+(A) = F$. A sub-base for a filter F is a set A such that I (a) is a base for F. If A is a sub-base for F, then $F = R_+(I(A))$, and we say that A generates F. Note that $A \subseteq I(a) \subseteq R_+(A)$.

Theorem 7.10 For each subset A of a Boolean B pre-algebra, the set $R_+(F(A))$ is a pre-filter. Any other pre-filter that contains A, contains $R_+(F(A))$. $R_+(F(A))$ is proper if and only if A has the pfi.

Proof Note that $x \in R_+(F(A))$ if and only if there $X \in \chi(A)$ such that if $y \in \text{Inf}(X)$, then $y \leq y$. Suppose that $x, y \in R_+(F(A))$. Then there are $X, Y \in \chi(A)$ such that if $x' \in$ Inf (X) and $y' \in$ Inf (Y), then $x' \leq x$ and $y' \in y$. It is clear that $X \cup Y \in \chi(A)$ and by Lemma 4.8, $\text{Inf}(X) \wedge \text{Inf}(Y) \in \text{Inf}(X \cap Y)$. But, $x' \leq y'$ and $x \leq y$ and therefore $x \wedge y \in R_+(I(A))$. It is immediate that if $x \in R_+(I(A))$ and $x \leq y$, then $y \in R_+(I(A))$. Therefore, $R_+(I(A))$ is a pre-filter.

Suppose that F is another filter that contains A and x is an element of $R_+(I(A))$. Therefore, there is $X \in \chi(A)$ such that $y \in$ Inf (X). So $y \leq x$. As $X \subseteq A \subseteq F$, it follows that Inf $(X) \subseteq F$ and therefore $x \subset F$.

Finally, suppose by the absurdity that there is $x \in \chi(A)$ such that Inf $(X) \subseteq [0]$. Therefore, $0 \in R_+(I(A))$ and, thus, $R_+(I(A))$ is a non-proprietary pre-filter. Suppose that $R_+(I(A))$ is not its pre-filter. So, for some $x \in \chi(A)$, $\text{Inf}(X) \subseteq [0]$ and, therefore, if $x \in \text{Inf}(X)$, then $x \equiv 0$. Thus, A does not have the pfi.

1.8 Ultra-Pre-filters

Definition 8.1 An ultra-prefilter of a pre-Boolean algebra B is a maximal proper pre-filter concerning the inclusion order of subsets of B.

Lemma 8.2 Let F be a pre-filter in a Boolean pre-algebra B. So F is an ultra-pre-filter if and only if for each $x \in$ B, or $x \in$ F, or $x' \in$ F, but not both.

Proof Suppose either $x \in$ F or $x' \in$ F. Let G be another pre-filter that properly contains F. Thus, there is $x \in$ G such that $x \in$ F. However, $x' \notin$ F by the hypothesis. Consequently, $x' \in$ G, which means that G = B.

Conversely, suppose that F is an ultra-pre-filter. If x is such that $x, x' \in$ F, $0 \equiv x \wedge x' \in$ F and therefore F would not be proper. Therefore, x and x' do not belong to F, for any $x \in$ B. Now let $x \in$ B be such that $x \in$ F. Let us show that $x' \in$ F. Consider the generated filter G of I ($\{x\}$ for this. Since F is maximal, it follows that G = B. Therefore, F ∪ $\{x\}$ does not have a pfi and, therefore, there is X ∈ χ(A) such that, if y ∈ Inf (X), $y \wedge x \equiv 0$. Thus, $y \leq x$ '. However, Inf (X) ⊆ F) means that $x' \in$ F.

Theorem 8.3 (Ultra-pre-filter theorem): Any ultra-pre-filter in a Boolean pre-algebra can be extended to an ultra-pre-filter.

Proof Let F be a pre-filter of B. Let us consider the set A of B's pre-filters that contain F. A can be partially ordered by including sets. A is not empty; therefore, F ∈ A. Let C* = $\{C_i \mid i \in I\}$ be a string in A. Consider the set C = $\cup_{i \in I} C_i$. We state that C ∈ A. In effect, let $x, y \in$ C. Therefore, there are i, j ∈ I such that $x \in C_i$ and $y \in C_j$. Since C* is a chain, it follows that $C_i \subseteq C_j$ or $C_j \subseteq C_i$. Suppose, without loss of generality, that $C_i \subseteq C_j$. Therefore, $x, y \in C_j$ and, so $x \wedge y \in C_j$. Consequently, $x \wedge y \in$ C. It is immediate that if $x \in C_j$, $z \in$ B and $x \leq z$., then $z \in$ C. Since $0 \notin C_j$, for all i ∈ I, it follows that 0 ∉ C and, therefore, C is its pre-filter. It is also immediate that F ⊆ C, since F ⊆ C_i, for all i ∈ I. Therefore, C ∈ A and, of course, C* is a upper bound for C in A. By Zorn's lemma, A contains a maximal element G, the ultra-pre-filter containing F.

Corollary 8.3.1 Any element not equivalent to 0 belongs to an ultra-pre-filter.

Proof Immediate.

Corollary 8.3.2 If $x, y \in$ B such that it is not the case that $x \equiv y$, then there is an ultra-pre-filter that contains exactly one of them.

Proof If it is not the case that $x \equiv y$, then it is not the case that $x \equiv y$ or it is not the case that $y \equiv x$. Suppose that it is not the case that $x \equiv$y. Thus, it is not the case that $x \wedge y' \equiv 0$, by Lemma 7.7. Therefore, the set $\{x, y'\}$ has the pfi and, therefore, there is an ultra-prefilter F such that $\{x, y'\} \subseteq$F. Like $y' \in$ F, it follows that $y \notin$ F (Lemma 8.2).

Corollary 8.3.3 Let A be a subset of a Boolean pre-algebra B such that $0 \notin A$. If for any two elements $x, y \in A$ we have $x \wedge y \in A$, then there is an ultra pre-filter containing A.

Proof Immediate.

1.9 Implicative (Absolute) Pre-lattices and Classic Implicative Pre-lattices

This paragraph defines the concepts of (absolute) implicative pre-lattice and classical pre-lattice.

Definition 9.1 A quadruple [R, $\equiv$, $\leq$, $\rightarrow$] is said to be an implicative pre-lattice (or absolute implicative pre-lattice) if [R, $\equiv$, $\leq$] is a pre-lattice and the $\rightarrow$ operation is such that:

(R8) $x \wedge (x \rightarrow y) \leq y$ for any $x, y \in$ R and.
(R9) $x \wedge y \leq z$ if and only if $x \leq y \rightarrow z$ for any $x, y, z \in$ R.

Definition 9.2 An implicative pre-lattice is said to be classic if the $\rightarrow$ operation satisfies the condition below:

(R10) $(x \rightarrow y) \rightarrow x \leq x$ for any $x, y \in$ R (Peirce's law).

Observation 9.3 From now on, we will leave aside the prefix "pre" of terms such as pre-algebra, pre-filter, etc. This observation will not lead to any ambiguity, as the meaning of the terminology will be apparent from the context.

References

1. Arruda, A.: A survey of paraconsistent logic. In: Arruda, A., Chuaqui, R., da Costa, N.C.A. (eds.) Mathematical Logic in Latin American, North-Holland, pp. 1–31 (1980)
2. Iséki, K.: Contribution to lattice theory. Publ. Math. Debrecen, vol. 12 (1951–1952), pp. 194–203
3. Smullyan, R.M.: First-Order Logic, Dover Publications, p. 176, ISBN-10: 0486683702, ISBN-13: 978-0486683706 (1995)
4. Iséki, K., Tanaka, S.: An introduction to the theory of BCK-algebras. Math. Japonica **23**, 1–26 (1978)
5. Halmos, P.H.: Algebraic Logic, Dover Publications, ISBN-10: 0486828042, ISBN-13: 978–0486828046 (2018)
6. Henkin, L., Monk, J.D., Tarski, A.: Cylindric Algebras, Part I, Lecture Notes in Mathematics (Book 883), p. 323, ISBN-13: 978-3540108818; ISBN-10: 3540108815 (1981)
7. Goldblatt, R.: Axiomatising the logic of computer programming. Lecture notes in computer science, vol. 130. Springer-Verlag, Berlin, Heidelberg, and New York (1982)
8. Jónsson B., Tarski A.: Boolean algebras with operators. Part I American Journal of Mathematics **73**(4), ISBN 891-10.2307/2372123 (1951)
9. Chagrov, A., Zakharyaschev, M.: Modal Logic. Clarendon Press (1997)
10. Priest, G.: An Introduction to Non-Classical Logic: From If to Is. Cambridge University Press (2002)
11. Hájek, P.: Metamathematics of Fuzzy Logic. Kluwer Academic Publishers (1998)
12. Johnstone, P.: Stone spaces. Cambridge University Press (1982)
13. Curry, H.B.: Leçons de Logique Algébrique, Gauthiers-Villars, p. 163 (1952)

14. Grätzer, G.: General lattice theory, Lehrbücher und Monographien aus dem Gebiete der exakten Wissenschaften/Mathematische Reihe (Book 52) Series on Pure and Applied Math, Birkhäuser Basel; 1 edn, p. 381, ISBN-10: 3764308133; ISBN-13: 978-3764308131 (1978)
15. Goldblatt, R.: Mathematical modal logic: A view of its evolution. J. Appl. Log. **4**(4), 377–402 (2006)
16. Henkin, L.: Le Structure Algèbrique des Théories Mathématiques, Gauthiers-Villars, 1st edn, Gauthier-Villars. 8vo., p. 52 (1956)
17. Sikorski, R.: Boolean Algebras. Springer-Verlag. ISBN-13: 978-3540044697, ISBN-10: 3540044698 (1969)
18. Curry, H.B.: Foundations of Mathematical Logic, Dover Books on Mathematics 2nd edn, p. 416, ISBN-10: 0486634620; ISBN-13: 978–0486634623 (2010)
19. Curry, H.B.: A theory of formal deducibility, 2nd edn, University of Notre Dame Press. In: Curry, H.B, Leçons de logique algébrique (in French). Paris: Gauthier-Villars (1952)
20. Curry, H.B.: Theory of formal deducibility, Univ of Notre Dame Pr, ISBN-10: 0268002746, ISBN-13: 978-0268002749 (1957)
21. Heyting, A.: Die formalen Regeln der intuitionistischen Logik. Sitzungsberichte der preußischen Akademie der Wissenschaften, phys.-math. Klasse (in German): 42–56, 57–71, 158–169 (1930)
22. Abe, J.M.: An introduction to the theory of Curry Systems. Sci. Math. Jpn **76**(2), 175–194, ISSN 1346-0862 (2013)
23. Abe, J.M., Nakamatsu, K.: Curry systems for algebraization of some non-classical logics. Int. J. Reason. based Intell. Syst. ISSN: 17550556, Interscience Enterprise, UK, pp. 59–69 (2011)
24. Abe, J.M., Nakamatsu, K., Akama, S.: Aspects of curry algebras, computability, constructibility, and topological spaces, frontiers in artificial intelligence and its applications, IOS Press, Amsterdam, Vol. 243, ISSN 0922-6389, pp. 1342–1349, (2012). https://doi.org/10.3233/978-1-61499-105-2-1342
25. Hintikka, J.: Form and content in quantification theory. Acta Philosophica Fennica **8**, 57–55 (1955)
26. da Costa, N.C.A.: Calculs propositionnels pour les systèmes formels inconsistants, C.R. Acad. Sc. Paris 257 (1963), pp. 3790–3793

Chapter 2
C_n-Algebras

The study of C_n-algebras is rooted in the algebraization of C_n-Calculus, a logical framework developed by the Brazilian logician Newton da Costa as part of his work on paraconsistent logics. C_n-Calculus, designed to tolerate contradictions without collapsing into triviality, introduced new ways of handling negation and contradiction within a formal logical system. However, when attempting to translate this calculus into an algebraic structure, it becomes evident that the negation operation in C_n-Calculus does not align with the fundamental equivalence relation typically expected between formulas in algebraic systems. This mismatch gives rise to a type of algebra known as C_n-algebras, which exhibit behaviors similar to Curry algebras; see da Costa [1–3].

In C_n-algebras, the behavior of negation requires a flexible, non-standard interpretation that diverges from classical algebraic expectations. This feature makes C_n-algebras a powerful tool for exploring the algebraic underpinnings of paraconsistent reasoning, where contradictions do not lead to trivialization but coexist within a structured, non-trivial logical framework.

The primary challenge in constructing C_n-algebras lies in adapting the negation operator to the context of algebraic equivalences. In classical logic and standard algebraic systems, negation is a well-behaved operation that aligns with equivalence relations, enabling consistent manipulation of formulas. In C_n-Calculus, however, negation is handled in a way that allows for the preservation of contradictory information, which means that the algebraic structure must accommodate situations where algebraically equivalent formulas do not behave identically under negation. This tension between equivalence and negation is a defining characteristic of C_n-algebras, and its resolution is key to understanding their algebraic properties.

In this chapter, we explore the construction and properties of C_n-algebras in detail, beginning with a formal definition of the algebraic structures that correspond to C_n-Calculus. We will examine how the standard operations of conjunction, disjunction, and implication are represented in C_n-algebras, focusing on how negation interacts with these operations. Additionally, we will investigate the meta-logical properties of

J. M. Abe, *An Introduction to Curry Systems*, Intelligent Systems Reference Library 280, https://doi.org/10.1007/978-3-031-97810-4_2

C_n-algebras, including consistency, soundness, and completeness, and their relationships to other known algebraic structures, such as de Morgan algebras and involutive lattices.

By establishing the algebraic foundations of C_n-Calculus through C_n-algebras, we aim to provide a rigorous mathematical framework for analyzing paraconsistent reasoning. This exploration enhances our understanding of the theoretical aspects of C_n-Calculus. It opens up new avenues for applying these algebras in fields that require managing inconsistent or contradictory information, such as information theory, legal reasoning, and artificial intelligence.

In this chapter, we study the *C*urry algebras—C_n-algebras, $1 \leq n \leq \omega$.

2.1 The C_n-Calculus

The C_n-Calculus, developed by Newton da Costa [4], represents one of the pioneering formal systems in paraconsistent logic, which seeks to address and manage contradictions without descending into triviality. Classical logic, by contrast, is governed by the Principle of Non-contradiction (*ex contradictione sequitur quodlibet*), which states that any proposition can be derived from a contradiction, making a system useless when faced with even a single contradiction. da Costa's C_n-Calculus offers a powerful alternative by rejecting this principle and allowing contradictions to coexist in a logically coherent framework without the system collapsing into inconsistency; see da Costa [4].

da Costa's C_n Calculus is part of a broader effort to develop a hierarchy of paraconsistent systems, labeled $C_1, C_2, \ldots, C_n$, where each system offers increasingly refined tools for handling contradictions. The hallmark of the C_n-Calculus is its ability to tolerate contradictions by carefully regulating the interaction of logical operations, particularly negation, in ways that do not lead to triviality.

At its core, the C_n-Calculus modifies key principles of classical logic, particularly emphasizing how negation is treated. In classical logic, negation is tightly bound to the Principle of Non-contradiction, which asserts that no statement can be both true and false simultaneously. In contrast, the C_n-Calculus weakens this law, allowing for both a formula and its negation to hold simultaneously in specific contexts without rendering the entire system trivial. This aspect is achieved by introducing new logical operators and axioms that control the propagation of contradictions; see da Costa [4]. In this way, one of the most significant features of the C_n-Calculus is its approach to negation.

Formally, the calculus can be understood as extending the usual syntax of propositional logic with an additional layer of control over how negation interacts with conjunctions and disjunctions.

da Costa introduced a hierarchy of systems, denoted C_1, C_2, ..., C_n, each with increasing expressive power and ability to manage contradictions. The lower systems, such as C_1, offer basic paraconsistent features by weakening the Principle of Non-contradiction and the Principle of Excluded Middle.

In these higher-level systems, more sophisticated rules govern the application of the consistency operator, allowing for a greater variety of contradictions to be handled while maintaining logical coherence. Each system in the hierarchy can be seen as refining the conditions under which contradictions can coexist with non-contradictory elements of the system, enabling a broader range of paraconsistent reasoning.

In essence, negation in the C_n-Calculus can be thought of as non-classical or even context-sensitive. The truth or falsity of a negated proposition does not automatically follow from the presence of its counterpart, nor does a contradiction immediately entail that all propositions become valid. This selective application of negation enables the system to handle inconsistent information without rendering it meaningless; see da Costa [4].

The C_n-Calculus has far-reaching implications, both theoretically and practically. Theoretically, it opens new possibilities for formalizing reasoning in contexts where contradictions are unavoidable or even expected. Philosophical and mathematical paradoxes, such as the Liar Paradox or the paradoxes of set theory (e.g., the existence of Russell's set), can be addressed within a paraconsistent framework without leading to triviality.

The C_n-Calculus is relevant in fields where incomplete or contradictory information is common, such as Artificial Intelligence, Information Theory, and Law. In Artificial Intelligence, systems often encounter conflicting data, and the ability to manage this data in a non-trivial way is essential for building robust models. Similarly, laws and precedents can sometimes be contradictory in legal reasoning, yet the legal system must still function coherently.

In summary, the C_n-Calculus by da Costa represents a significant advance in the study of non-classical logics, pois os mesmos são os primeiros sistemas importantes na literatura.

In this paragraph, we will investigate these systems, addressing their most important properties with a view to their algebraization. In this paragraph, we begin by describing the propositional calculus C_1.

The primitive symbols of C_1 are as follows: (1) an enumerable set of propositional variables; (2) the basic connectives of the calculus are: $\rightarrow$ (if, … then …), $\wedge$ (and), $\vee$ (or) and $\neg$ (negation); the symbol $\leftrightarrow$ (if, and only if) is defined as usual; (3) auxiliary symbols: parentheses.

The concept of the formula is defined in the usual way, as well as the other conventional syntactic notions such as proof, syntactic consequence ($\vdash$), etc.

Intuitively, in C_1 we want the connectives $\rightarrow$, $\wedge$, $\vee$ to "behave" as in classical logic. Propositions are subdivided into two categories: those that satisfy the Principle of Non-contradiction, $\neg(A \wedge \neg A)$, called "well-behaved" propositions, and those that do not (called "badly behaved" propositions). For "well-behaved" propositions, the principles of classical logic apply, while for "badly behaved" propositions, they and their negations are true. Calculus C_1, too, is constructed so that any proposition formed from "well-behaved" propositions through logical connectives is also "well-behaved.“

The postulates of C_1 (axiom schemes and Inference rule) are as follows, where A, B, and C are formulas and A^o is short for $\neg(A \wedge \neg A)$:

($\rightarrow_1$) $\rightarrow_1)A \rightarrow (B \rightarrow A)$
($\rightarrow_2$) $(A \rightarrow B) \rightarrow ((A \rightarrow (B \rightarrow C)) \rightarrow (A \rightarrow C))$
($\rightarrow_3$) $\frac{A, A \rightarrow B}{B}$(MP (Modus Ponens))
($\wedge_1$) $A \wedge B \rightarrow A$
($\wedge_2$) $A \wedge B \rightarrow B$
($\wedge_3$) $A \rightarrow (B \rightarrow (A \wedge B))$
($\vee_1$) $A \rightarrow (A \vee B)$
($\vee_2$) $B \rightarrow (A \vee B)$
($\vee_3$) $(A \rightarrow C) \rightarrow ((B \rightarrow C) \rightarrow (A \vee B \rightarrow C))$
($\neg_1$) $\neg\neg A \rightarrow A$
($\neg_2$) $A \vee \neg A$
($\neg_3$) $B \rightarrow ((A \rightarrow B) \rightarrow ((A \rightarrow \neg B) \rightarrow \neg A))$
($\neg_4$) $A^o \wedge B^o \rightarrow (A \rightarrow B)^o \wedge (A \wedge B)^o \wedge (A \vee B)^o$
($\neg_5$) $A^o \rightarrow (\neg A)^o$

The postulates $\rightarrow_1$ - $\rightarrow_3$, $\wedge_1$ - $\wedge_3$, and $\vee_1$ - $\vee_3$ imply that all the properties of positive intuitionist logic are valid in C_1. Peirce's law is valid in C_1 and, thus, the connectives $\rightarrow$, $\wedge$ and $\vee$ satisfy the postulates of classical positive logic. The postulate $\neg_1$ implies that any formula A, or A is "well behaved"or A is "badly behaved". $\neg_2$ is a corollary of the requirement that any formula is true or false. $\neg_3$ is a weak form of the principle of reduction to absurdity, and finally, $\neg_4$ tells us that any formula obtained from "well-behaved" formulas is "well-behaved," $\neg_5$ tells us that a well-behaved proposition entails a well-behaved negation.

Let us define the operator $\neg^*$ (called a strong negation): if A is a formula, $\neg^* A$ abbreviates $\neg A \wedge A^o$. In C_1, the connectives $\rightarrow$, $\wedge$, $\vee$, $\neg^*$ and $\leftrightarrow$ have all the classic properties of implication, conjunction, disjunction, negation, and equivalence, respectively. Thus, the classical propositional calculus is 'contained' in C_1, although this calculus is strictly weaker than classical propositional logic. Furthermore, the following conditions can be proved:

(a) The Principle of Non-contradiction, in the form $\neg(A \wedge \neg A)$, is not generally valid.
(b) We cannot deduce any formula B from two contradictory premises A and $\neg A$.
(c) C_1 contains some schemes and rules of classical logic compatible with the two preceding conditions.

We now move on to the study of C_n systems.

Definition 1.1 Let A be a formula. We define A inductively by the clauses:

(a) A^1 abbreviates A^o
(b) A^n abbreviates $(A^{n-1})^o$

Definition 1.2 Let A be a formula. We define $A^{(n)}$ inductively as:

(a) $A^{(0)}$ abbreviates A
(b) $A^{(1)}$ abbreviates A^o
(c) $A^{(n)}$ abbreviates $A^{(n-1)} \wedge (A^{(n-1)})^o$

The postulates of C_n, $1 \leq n < \omega$, are the postulates $\rightarrow_1$ to $\vee_3$ of C_1 plus the following:

($\neg_{n1}$) $B^{(n)} \rightarrow ((A \rightarrow B) \rightarrow ((A \rightarrow \neg B) \rightarrow \neg A))$
($\neg_{n2}$) $A^{(n)} \wedge B^{(n)} \rightarrow (A \rightarrow B)^{(n)} \wedge (A \wedge B)^{(n)} \wedge (A \vee B)^{(n)}$
($\neg_{n3}$) $A^{(n)} \rightarrow (\neg A)^{(n)}$

The postulates of C_ω are the postulates $\rightarrow_1$ to $\neg_2$ of C_1.

The calculus C_n, $1 \leq n < \omega$, has many of the properties of classical propositional calculus. Thus, the deduction theorem, the rule of proof by cases, and the rule of double negation are valid.

Below, we will demonstrate some C_1 results necessary for subsequent developments.

Lemma 1.3 For every formula A of C_1, we have $\vdash A \rightarrow A$.

Proof Exercise.

Lemma 1.4 For any formulas A, B of C_1, we have $A, B \vdash A \wedge B$.

Proof Exercise.

Lemma 1.5 For any formulas A, B of C_1, we have $A \wedge B \vdash B \wedge A$.

Proof Exercise.

Lemma 1.6 If A and B are formulas of C_1, we have $\vdash A \wedge B$ and $\vdash B \wedge C$ entails $\vdash A \wedge C$.

Proof Exercise.

Lemma 1.7 If A, B, and C are formulas such that $\vdash A \rightarrow B$ and $\vdash B \rightarrow C$, then $\vdash A \rightarrow C$.

Proof Exercise.

Lemma 1.8 If A, B, and C are formulas, then $\vdash (A \rightarrow ((B \rightarrow A) \rightarrow C)) \rightarrow (A \rightarrow C)$.

Proof Exercise.

Lemma 1.9 If A, B, and C are formulas such that $\vdash A \to B$, then $\vdash (A \vee C) \to (B \vee C)$.

Proof Exercise.

Lemma 1.10 If $\vdash A \to B$, then $A \to B \vdash A \wedge C \to B \wedge C$.

Proof Exercise.

Lemma 1.11 If A and B are formulas, then $A, B \vdash A \to B$.

Proof Exercise.

Lemma 1.12 If A, B are formulas, from $\vdash A \leftrightarrow B$ it cannot be concluded that $\vdash \neg A \leftrightarrow \neg B$.

Proof Exercise.

2.2 The C_1-Algebra

Next, the algebraic systems we will study algebraize the various systems C_n, $1 \leq n < \omega$, of da Costa. The direct result of these studies is that C_n-algebras, $1 \leq n < \omega$. They are also an interesting generalization of Boolean algebras [5, 6].

Let us first study C_1-algebras.

Definition 2.1 A C_1-algebra (or C_1-pre-lattice) is an absolute implicative pre-lattice [S, ≡, →, ∧, ∨, ’] with a maximum element 1 and with an operator’, satisfying the following conditions, where $x^o = (x \wedge x')'$ and $x \equiv y \Leftrightarrow x \leq y$ and $y \leq x$:

(S_1) $x \vee x' \equiv 1$.
(S_2) $x'' \leq x$.
(S_3) $y^o \leq (x \to y) \to ((x \to y') \to x')$.
(S_4) $x^o \wedge y^o \leq (x \to y)^o$
(S_5) $x^o \wedge y^o \leq (x \wedge y)^o$
(S_6) $x^o \wedge y^o \leq (x \vee y)^o$
(S_7) $x^o \leq (x')^o$

Consider the C_1 calculus. Let F be the class of C_1 formulas. Let us define operations →, ∧, ∨ and ’ in F, respectively, by:

1. $x \to y =_{\text{Def}} x \to y$
2. $x \wedge y =_{\text{Def}} x \wedge y$
3. $x \vee y =_{\text{Def}} x \vee y$
4. $x' =_{\text{Def}} \neg x$

The $\to$, $\wedge$ and ' symbols of the first member of the above definition denote the C_1-algebra's operations, and the second member denotes the logical connectives of C_1.

Let us define the relation $\equiv$ over F like this:

$x \equiv y$ if $\vdash x \leftrightarrow y$.

We have the following important theorem:

Theorem 2.2 With the notations and definitions above, the algebraic system [F, $\equiv$, $\wedge$, $\vee$] has a pre-lattice structure.

Proof First, let us verify that $\equiv$ is an equivalence relation.

(1) $x \equiv x$. For that, we need to prove that $\vdash x \leftrightarrow x$, or, $\vdash (x \to x) \wedge (x \to x)$.

1. $\vdash x \leftrightarrow x$ Lemma 1.3
2. $\vdash (x \to x) \wedge (x \to x)$ Lemma 1.4, thus, $x \equiv x$.

(2) Suppose that $x \equiv y$, or, $\vdash x \leftrightarrow y$.

1. $\vdash (x \to y) \wedge (y \to x)$ Hypotesis
2. $\vdash (y \to x) \wedge (x \to y)$ Lemma 1.5
3. $\vdash (y \leftrightarrow x)$, thus $y \equiv x$

(3) Suppose that $x \equiv y$ and $y \equiv z$, or , $\vdash x \leftrightarrow y$ and $\vdash y \leftrightarrow z$.

1. $\vdash (x \to y) \wedge (y \to x)$ Hypotesis
2. $\vdash (y \to z) \wedge (z \to y)$ Hypotesis
3. $\vdash (x \to y)$ and $\vdash (y \to x)$ 1,2, $\wedge_1$
4. $\vdash (x \to z)$ Lemma 1.7
5. $\vdash (z \to y)$ and $\vdash (y \to x)$ 1, 2, $\wedge_2$
6. $\vdash (z \to x)$ Lemma 1.7
7. $\vdash (x \to z) \wedge (z \to x)$ Lemma 1.7
8. $\vdash (x \leftrightarrow z)$ and thus, $x \equiv z$

It follows, therefore, that $\equiv$ is an equivalence relation.

Let us now verify that R is pre-lattice.

(L_1) $x \vee y \equiv y \vee x$.

1. $(x \to (y \vee x)) \to ((y \to (y \vee x)) \to ((x \vee y) \to (y \vee x)))$ $\vee_3$
2. $x \to y \vee x$ $\vee_2$
3. $(y \to (y \vee x)) \to ((x \vee y) \to (y \vee x))$ 1,2, MP
4. $y \to (y \vee x)$ $\vee_1$
5. $(x \vee y) \to (y \vee x)$ 3,4, MP

Similarly, we have.

6. $(y \vee x) \to (x \vee y)$
7. $(x \vee y) \to (y \vee x) \wedge (y \vee x) \to (x \vee y)$ Lemma 1.4
8. $(x \vee y) \to (y \vee x)$, which entails $x \vee y \equiv y \vee x$

(L_2) $x \vee x \equiv x$.

1.	$x \to (x \to x)$	$\vee_1$
2.	$(x \to x) \to ((x \to x) \to (x \vee x \to x)$	$\vee_3$
3.	$(x \to x)$	Lemma 1.3
4.	$x \vee x \to x$	3, 2, MP
5.	$(x \vee x \to x) \wedge (x \to (x \vee x))$	1, 4, Lemma 1.4
6.	$x \vee x \to x$, or, $x \vee x \equiv x$	

(L_3) $(x \vee y) \vee z \equiv x \vee (y \vee z)$.

1.	$(x \to ((x \vee y) \vee z)) \to (((y \vee z) \to ((x \vee y) \vee z)) \to (x \vee (y \vee z) \to ((x \vee y) \vee z)))$ $\vee_3$	
2.	$x \to (x \vee y)$	$\vee_1$
3.	$x \vee y \to ((x \vee y) \vee z)$	$\vee_1$
4.	$x \to ((x \vee y) \vee z)$	2, 3, Lemma 1.7
5.	$((y \vee z) \to ((x \vee y) \vee z)) \to (x \vee (y \vee z) \to ((x \vee y) \vee z))$	1, 4, MP
6.	$y \to (x \vee y)$	$\vee_2$
7.	$x \vee z \to ((x \vee y) \vee z)$	6, Lemma 1.9
8.	$x \vee (y \vee z) \to (x \vee y) \vee z$	7, 5, MP

We now prove that $(x \vee y) \vee z \to x \vee (y \vee z)$

9.	$((x \vee y) \to ((x \vee y) \vee z) \to ((z \to x \vee y))) \to ((x \vee y) \vee z \to x \vee (y \vee z))$ 9, 11, MP	
10.	$y \to (y \vee z)$	$\vee_1$
11.	$x \vee y \to x \vee (y \vee z)$	$\vee_2$
12.	$z \to (x \vee (y \vee z)) \to ((x \vee y) \vee z \to x \vee (y \vee z))$	9, 11, MP
13.	$z \to (y \vee z)$	$\vee_2$
14.	$y \vee z \to (x \vee (y \vee z))$	$\vee_2$
15.	$.z \to x \vee (y \vee z)$	13, 14, Lemma 1.7
16.	$(x \vee y) \vee z \to x \vee (y \vee z)$	15, 12, MP
1.	Therefore, it follows that	
17.	$((x \vee y) \vee z \to x \vee (y \vee z)) \wedge (x \vee (y \vee z) \to (x \vee y) \vee z)$	
18.	$((x \vee y) \vee z \to x \vee (y \vee z)) \leftrightarrow (x \vee (y \vee z) \to (x \vee y) \vee z)$, or,	

$(x \vee y) \quad \vee z \equiv x \vee (y \vee z)$

(L_4) $\quad x \wedge y \equiv y \wedge x$

1.	$x \wedge y$	Hypotesis
2.	$x \wedge y \to x$	$\wedge_1$
3.	$x \wedge y \to y$	$\wedge_2$
4.	x, y	1, 2, 3, MP
5.	$y \to (x \to y \wedge x)$	$\wedge_3$
6.	$y \wedge x$	4, 5, MP
7.	$x \wedge y \to y \wedge x$	Deduction Theorem

Similarly, we have

8. $y \wedge x$
9. $(x \wedge y \to y \wedge x) \wedge (y \wedge x \to x \wedge y)$ Lemma 1.4
10. $x \wedge y \leftrightarrow y \wedge x$,

That is, $x \wedge y \equiv y \wedge x$

(L_5) $x \wedge x \equiv y$

1.	$x \wedge x \to x$	$\wedge_1$
2.	x	Hypotesis
3.	$x \to (x \to x \wedge x)$	$\wedge_3$
4.	$x \wedge x$	2, 3, MP
5.	$x \wedge x \to x$	2, 4, Deduction theorem
6.	$(x \wedge x \to x) \wedge (x \to x \wedge x)$	1, 5, Lemma 1.4
7.	$x \wedge x \leftrightarrow x$, or , $x \wedge x \equiv x$	

(L_6) $(x \wedge y) \wedge z \equiv x \wedge (y \wedge z)$

1.	$(x \wedge y) \wedge z$	Hypotesis
2.	$(x \wedge y) \wedge z \to z$	$\wedge_2$
3.	z	1, 2, MP
4.	$x\ (x \wedge y) \wedge z \equiv x \wedge y$	$\wedge_1$
5.	$x \wedge y$	1, 4, MP
6.	$x \wedge y \to x$	$\wedge_1$
7.	$x \wedge y \to y$	$\wedge_1$
8.	x, y	6, 7, MP
9.	$x \to (y \wedge z \to x \wedge (y \wedge z))$	$\wedge_3$
10.	$(y \wedge z) \to x \wedge (y \wedge z)$	8, 9, MP
11.	$y \to (z \to (y \wedge z))$	$\wedge_3$
12.	$y \wedge z$	3, 7, MP
13.	$x \wedge (y \wedge z)$	10, 12, MP
14.	$(x \wedge y) \wedge z \to x \wedge (y \wedge z)$	1, 13, Deduction Theorem

We also have that:

15.	$(x \wedge y) \wedge z$	Hypotesis
16.	$x \wedge (y \wedge z) \to x$	$\wedge_1$
17.	x	15, 16, MP
18.	$x \wedge (y \wedge z) \to y \wedge z$	$\wedge_2$
19.	$y \wedge z$	15, 18, MP
20.	$y \wedge z \to y, z$	$\wedge_1, \wedge_2$
21.	y, z	19, 20, MP
22.	$(x \wedge y) \to (z \to (x \wedge y) \wedge z)$	$\wedge_3$
23.	$x \to (y \to x \wedge y)$	$\wedge_3$
24.	$x \wedge y$	17, 21, MP
25.	$z \to (x \wedge y) \wedge z$	22, 24, MP
26.	$(x \wedge y) \wedge z$	21, 25, MP

27. $x \wedge (y \wedge z) \to (x \wedge y) \wedge z$ — 15, 26, Deduction Theorem
28. $((x \wedge y) \wedge z \to x \wedge (y \wedge z)) \wedge (x \wedge (y \wedge z) \to (x \wedge y) \wedge z)$ — 14, 27, Lemma 1.4
29. $(x \wedge y) \wedge z \leftrightarrow x \wedge (y \wedge z)$, or,
30. $(x \wedge y) \wedge z \equiv x \wedge (y \wedge z)$

(L_7) $x \wedge (x \vee y) \equiv x$

1. $x \wedge (x \vee y)$ — Hypotesis
2. $x \wedge (x \vee y)$ — $\wedge_1$
3. x — 1, 2, MP
4. $x \wedge (x \vee y) \to x$ — 1, 3, Deduction theorem

We also have:

5. x — Hypotesis
6. $x \to ((x \vee y) \to x \wedge (x \vee y))$ — $\wedge_3$
7. $x \vee y \to x \wedge (x \vee y)$ — 5, 6, MP
8. $x \to x \vee y$ — $\vee_1$
9. $x \vee y$ — 5, 8, MP
10. $x \wedge (x \vee y)$ — 7, 9 MP
11. $x \to (x \wedge (x \vee y))$ — 5, 10, Deduction Theorem
12. $((x \wedge (x \vee y) \to x) \wedge (x \to (x \wedge (x \vee y))$ — 4, 11, Lemma 1.4
13. $x \wedge (x \vee y) \leftrightarrow x$, or $x \wedge (x \vee y) \equiv x$

(L_8) $x \vee (x \wedge y) \equiv x$

1. $x \vee (x \wedge y)$ — Hypotesis
2. $(x \to x) \to [((x \wedge y) \to x) \to ((x \vee (x \wedge y)) \to x)]$ — $\vee_3$
3. $x \to x$ — Lemma 1.3
4. $(x \wedge y \to x) \to ((x \vee (x \wedge y)) \to x)$ — 2, 3 MP
5. $x \wedge y \to x$ — $\wedge_1$
6. $x \vee (x \wedge y) \to x$ — 4, 5 MP

We also have that:

9. x — Hypotesis
10. $x \to x \vee (x \wedge y)$ — $\vee_2$
11. $x \vee (x \wedge y) \to x$ and $x \to x \vee (x \wedge y)$, therefore, $x \vee (x \wedge y) \leftrightarrow x$.

Thus, $x \vee (x \wedge y) \equiv x$.
(L_9). If $x \equiv x'$ and $y \equiv y'$, then $x \wedge y \equiv x' \wedge y'$ and $x \vee y \equiv x' \vee y'$.

(a) $x \wedge y \leftrightarrow x' \wedge y'$
1. $x \wedge y$ — Hypotesis
2. $x \to x' \wedge x' \to x$ — Hypotesis
3. $y \to y' \wedge y' \to y$ — Hypotesis
4. $(x \to x' \wedge x' \to x) \to (x \to x')$ — $\wedge_1$
5. $x \to x'$ — 2, 4 MP

6. $x \wedge y \rightarrow x' \wedge y$ — 5, Lemma 1.10
7. $(y \rightarrow y' \wedge y' \rightarrow y) \rightarrow (y \rightarrow y')$ — $\wedge_1$
8. $y \rightarrow y'$ — 3, 7 MP
9. $x' \wedge y \rightarrow x' \wedge y'$ — Lemma 1.10
10. $x \wedge y \rightarrow x' \wedge y'$ — 6, 9, Lemma 1.7

Similarly, we have

11. $x' \wedge y' \rightarrow x \wedge y$
12. $(x \wedge y \rightarrow x' \wedge y') \wedge (x' \wedge y' \rightarrow x \wedge y)$ — 10, 11, Lemma 1.4
13. $x \wedge y \leftrightarrow x' \wedge y'$ and, therefore, $x \wedge y \equiv x' \wedge y'$

(b) $x \vee y \equiv x' \vee y'$

14. $x \rightarrow x'$ — 5
15. $x \vee y \rightarrow x' \vee y$ — 14, Lemma 1.9
16. $y \rightarrow y'$ — 8
17. $x' \vee y \rightarrow x' \vee y'$ — 16, Lemma 1.9
18. $x \vee y \rightarrow x' \vee y$ — 15, 17, Lemma 1.7

Similarly, we have

19. $x' \vee y' \rightarrow x \vee y$
20. $(x \vee y \rightarrow x' \vee y') \wedge (x' \vee y' \rightarrow x \vee y)$ — 18, 19, Lemma 1.4
21. $x \vee y \leftrightarrow x' \vee y'$ and, therefore, $x \vee y \equiv x' \vee y'$

Therefore [F, $\equiv$, $\rightarrow$, $\wedge$, $\vee$] has a pre-lattice structure.

Lemma 2.3 With the notations of the previous theorem, we have:

(1) $x \equiv y$ if and only if $x \leq y$ and $y \leq x$, where $x \leq y =_{\text{Def}} x \equiv x \wedge y$.
(2) $x \leq y$ if and only if $x \vdash y$.

Proof (1). It is an immediate consequence of Corollary 4.7.1.

(2) Suppose that $x \leq y$, that is, $x \equiv x \wedge y$, which is equivalent to stating that $\vdash x \leftrightarrow x \wedge y$.

1. x — Hypotesis
2. $x \leftrightarrow x \wedge y$ — Hypotesis
3. $(x \rightarrow x \wedge y) \wedge (x \wedge y \rightarrow x)$ — 2
4. $(x \rightarrow x \wedge y) \wedge (x \wedge y \rightarrow x) \rightarrow (x \rightarrow x \wedge y)$ — $\wedge_1$
5. $x \rightarrow x \wedge y$ — 3, 4 MP
6. $x \wedge y$ — 1, 5, MP
7. $x \wedge y \rightarrow y$ — $\wedge_2$
8. y — 6, 7, MP

Thus, $x \vdash y$.

Conversely, suppose that $x \vdash y$.

1. x — Hypotesis
2. y — Hypotesis

3. $x \wedge y$ — 1, 2, Lemma 1,4
4. $x \to x \wedge y$ — 1, 3, Lemma 1,11
5. $x \wedge y \to x$ — $\wedge_1$
6. $(x \to x \wedge y) \wedge (x \wedge y \to x)$ — 4, 5, Lemma 1.4
7. $x \leftrightarrow x \wedge y$, that is, $x \equiv x \wedge y$. Therefore, $x \leq y$.

Theorem 2.4 With the notations of the previous theorem, the pre-lattice [F, $\equiv$, $\to$, $\wedge$, $\vee$] satisfies the axioms (R_8) and (R_9), that is, [F, $\equiv$, $\to$, $\wedge$, $\vee$] is an absolute implicative pre-lattice.

Proof (R_8) $x \wedge (x \to y) \leq y$, for any $x, y \in$ F.

1. $x \wedge (x \to y)$ — Hypotesis
2. $x \wedge (x \to y) \to x$ — $\wedge_1$
3. x — 1, 2, MP
4. $x \wedge (x \to y) \to (x \to y)$ — $\wedge_2$
5. $x \wedge (x \to y) \leq y$
6. y — 3, 5, MP

It follows, therefore, that $x \wedge (x \to y) \vdash y$, that is,
(R_9) $x \wedge y \leq z$ if and only if $x \wedge y \leq z$.
a) Let us assume that $x \wedge y \leq z$.

1. x, y — Hypotesis
2. $x \to (y \to x \wedge y)$ — $\wedge_3$
3. $x \wedge y$ — 1, 2, MP
4. z — 3, Hypotesis

We then have $x, y \vdash z$ and therefore $x \vdash y \to z$ (Deduction theorem). Therefore, $x \leq y \to z$.

b) Conversely, suppose that $x \leq y \to z$.

1. $x \wedge y$ — Hypothesis
2. $x \wedge y \to x, y$ — $\wedge_1$, $\wedge_2$
3. x, y — 1, 2, MP
4. $y \to z$ — 3, Hypothesis
5. z — 4, 3 MP

It follows that $x, y \vdash z$, that is, $x \wedge y \leq z$.

Lemma 2.5 The operator ' defined above satisfies conditions S1 - S7 of definition 2.1.

Proof It is a consequence of the axioms $\neg_1$ - $\neg_4$ of calculus C_1.

Theorem 2.6 With the notations of the preceding theorems, the system [F, $\equiv$, $\to$, $\wedge$, $\vee$, '] constitutes a C_1-algebra.

Proof Due to theorems 2.2, 2.4 and Lemmas 2.3, 2.5.

Theorem 2.7 Let [F, ≡, →, ∧, ∨, ’] be a C_1-algebra. The operator’ is not monotonous regarding the fundamental relation ≡.

Proof See Mortensen [7]

Corollary 2.7.1 Every C_1-algebra is a Curry algebra.

Proof Due to the previous theorem.

Theorem 2.8 All C_1-algebra is a classic implicative pre-lattice. Therefore, for any x, y elements of this algebra, the following relations are valid:

(1) $x \vee (x \to y) \equiv 1$
(2) $(x \to y) \to x$ (Peirce’s law)
(3) $(x \to y) \to x \vee y$

Proof Immediate.

Theorem 2.9 In a C_1-algebra, the following expressions are valid for all x, y, and z elements of that algebra:

(1) If $x \leq y$, then $x \to z \leq y \to z$
(2) If $x \leq y$, then $z \to x \leq z \to y$
(3) $y \leq x \to y$
(4) $x \to (y \to z) \equiv x \wedge y \to z \equiv y \to (x \to z)$
(5) $x \wedge (x \to y) \equiv x \wedge y$
(6) $x \to (y \to z) \leq (x \to y) \to (x \to z)$
(7) $x \to y \leq (x \to y) \wedge (x \to z)$
(8) $x \to y \leq x \wedge z \to y \wedge z$
(9) $(x \to y) \wedge (y \to z) \leq (x \to z)$
(10) $x \to x \wedge y \to x \to y \equiv x \to (x \to y)$
(11) $(x \to y) \vee (x \to z) \leq x \to (y \vee z)$
(12) $(x \to z) \vee (y \to z) \leq x \wedge y \to z$
(13) $z \vee (x \to y) \leq (x \vee z) \to (y \vee z)$
(14) $(x \vee y) \wedge (x \to y) \equiv y$
(15) $(x \vee y) \leq (x \to y) \to y$
(16) $(x \to y) \wedge (y \to x) \equiv (x \vee y) \to (x \wedge y)$

Theorem 2.10 In a C_1-algebra, there is a maximum element, and we have:

(1) $x \wedge (x \to y) \leq y$
(2) If $x \wedge z \leq y$, then $z \leq x \to y$
(3) $x \equiv 1 \to x$
(4) $x \leq y$ if and only if $1 \leq x \to y$
(5) $(x \vee y) \to z \equiv (x \to z) \vee (y \to z)$

Proof Analogous to the previous theorem.

Theorem 2.11 All C_1-algebra are distributive.

Proof Immediate.

Theorem 2.12 Let R be a C_1-algebra. Any element x of R, x * abbreviates $x' \wedge x^o$.

Theorem 2.13 In a C_1-algebra, x * is the Boolean complement of x; therefore, $x \vee x^* \equiv 1$ and $x \wedge x^* \equiv 0$.

Proof We initially prove that $x \vee x^* \equiv 1$. Indeed,

1. $x^o \leq x \vee x^o$
2. $(x^o)' \leq ((x \wedge x')')' \leq x \wedge x' \leq x \leq x \vee x^o$
3. $x^o \vee (x^o)' \leq x \vee x^o$ 1, 2
4. $x \vee (x^o)' \equiv 1$
5. $x \vee x^o \equiv 1$
6. $x \vee x' \equiv 1$
7. $(x \vee x') \wedge (x \vee x^o) \equiv 1$ 3, 4
8. $x \vee (x' \wedge x^o) \equiv 1$
9. $x \vee x^* \equiv 1$

We now prove that $x \wedge x^* \equiv 0$,

1. $x \wedge x^* \equiv x \wedge x' \wedge x^o$
2. $x \wedge x^* \wedge 1 \leq x$
3. $x \wedge x^* \leq 1 \to x$
4. $x \wedge x^* \wedge 1 \leq x'$
5. $x \wedge x^* \leq 1 \to x'$
6. $x \wedge x^* \wedge 1 \leq x^o$
7. $x \wedge x^* \leq 1 \to x^o$
8. $x \wedge x^* \leq (1 \to x) \wedge (1 \to x') \wedge (1 \to x^o)$
9. $x \wedge x^* \leq 1 \to [x^o \to ((1 \to x) \to (1 \to x'))]$
10. $x \wedge x^* \leq 1 \to 1'$
11. $x \wedge x^* \leq 1'$
12. $x \wedge x^* \leq 0$
13. $x \wedge x^* \equiv 0$

Theorem 2.14 In a C_1-algebra, the structure of the set underlying this pre-lattice and the operations $\wedge$, $\vee$, * is a Boolean pre-algebra.

Proof It follows from the previous theorem.

Observation 2.15 According to Theorem 2.13, the structure of the set underlying a C_1-algebra and the operations $\wedge$, $\vee$, * is a Boolean pre-algebra, whose fundamental relation is the equivalence relation $\equiv$. Under Theorem 6.3 (Chap. I) passing to the quotient, we obtain a Boolean algebra in the usual sense. Based on this observation, we formulate the following:

Definition 2.16 Let R = [R, ≡, → , ∧, ∨, ’] be a C_1-algebra and R/≡ the Boolean algebra obtained by passing the Boolean pre-algebra R to the quotient. Any Boolean algebra isomorphic to the R/≡ algebra is called Boolean algebra associated with the C_1-algebra.

The theory of representation of Stone’s Boolean algebras, whether from the set-theoretical prism or topological point of view, can be extended to C_1-algebras. The most relevant results regarding the representation of C_1-algebras are as follows:

Theorem 2.17 Every C_1-algebra is associated with a reduced field of sets.

Proof Due to the above observation, any C_1-algebra is associated with a Boolean algebra. By Stone’s representation theorem, all Boolean algebras are isomorphic to a reduced field of sets.

Theorem 2.18 All C_1-algebra is associated with the field of sets simultaneously open and closed in a compact and totally disconnected Hausdorff space [3, 8].

Proof Similar to the previous theorem.

Problem 2.19 How many non-isomorphic C_1-algebras are associated with the same Boolean algebra?

2.3 The C_n-Algebras, $1 \leq n < \omega$

Next, we will study the algebraic systems and algebraize the different C_n, $1 \leq n < \omega$, systems by da Costa; see da Costa [4]. The direct result of these studies is that C_n-algebras, $1 \leq n < \omega$, are also an interesting generalization of Boolean algebras.

All concepts and results related to C_1-algebra extend, with adaptations, to C_n-algebras, $1 \leq n < \omega$. We mention, however, some results.

Definition 3.1 Let [S, → , ∧, ∨] be a C_n-algebra. If $x \in S$, x^1 abbreviates x^o. x^n abbreviates $x^{oo\ldots o}$, where the symbol ^{o} appears n times in this last expression.

Definition 3.2 Under the conditions of the preceding definition, $x^{(1)}$ abreviates x^1, $x^{(n+1)}$ abreviates $x^{(n)} \wedge x^{n+1}$.

Definition 3.3. A C_n-algebra (or C_n-pre-lattice) is an absolute implicative pre-lattice [S, ≡, →, ∧, ∨, ’] with a maximum element 1 and an operator’, satisfying the following conditions:

(S’$_1$) $x \vee x' \equiv 1$.

(S’$_2$) $x'' \leq x$.

(S’$_3$) $y^{(n)} \leq (x \to y) \to ((x \to y') \to x')$.

(S’$_4$) $x^{(n)} \wedge y^{(n)} \leq (x \to y)^{(n)}$

(S’$_5$) $x^{(n)} \wedge y^{(n} \leq (x \wedge y)^{(n)}$

(S’$_6$) $x^{(n)} \wedge y^{(n} \leq (x \vee y)^{(n)}$

(S’$_7$) $x^{(n)} \leq (x')^{(n)}$

Theorem 3.4 Every $C_{n,}$ $1 \leq n < \omega$, is a classical and distributive implicative pre-lattice.

Proof Analogous to Theorems 2.8 and 2.11.

2.4 The C_ω-Algebra

Definition 4.1 A C_ω-algebra is an absolute implicative pre-lattice [S, ≡, →, ∧, ∨, ’] with first element 1, in which an operator ’ is defined as satisfying the conditions:

(S_1) $x \vee x' \equiv 1$
(S_2) $x'' \leq x$

Consider the propositional calculus C_ω. Let F be the class of the formulas of $C_{\omega,}$ →, ∧, ∨, ’ the operations on F and ≡ the relation on F as in §2.

Theorem 4.2 The system [F, ≡, →, ∧, ∨, ’] is a C_ω-algebra, that is, C_ω-algebra is an algebraic version of C_ω Calculus.

Proof Similar to Theorem 2.6.

Example 4.3 The example presented below is a C_ω-algebra. It is a Curry algebra since the operator ’ is not monotonic concerning the fundamental equivalence relation between formulas [9].

Consider, then, the matrix M = [{1, 2, 3}, {1, 2}, → , ∧, ∨, ¬], where

$A \to B$:

A/*B*	1	2	3
1	1	1	3
2	1	1	3
3	1	1	1

$A \wedge B$:

A/*B*	1	2	3
1	1	1	3
2	1	1	3
3	3	3	3

$A \vee B$:

A/B	1	2	3
1	1	1	1
2	1	1	1
3	1	1	3

$\neg A$:

A	$\neg A$
1	3
2	2
3	1

It is easy to prove that it is a C_ω-algebra and that the $\neg$ operator is not compatible with the fundamental equivalence relation between formulas.

Theorem 4.4 A C_ω-algebra has, in general, no smallest element.

Proof It follows from the fact that the C_ω calculus is infinitely trivializable. This can be seen considering the matrix M = [N, I, →, ∧, ∨, ¬], where N is the set of positive natural numbers, I is the set of odd numbers and operations →, ∧, ∨, ¬ are defined as follows (*v*, *vd*, *vnd* are, respectively, abbreviations for 'value', 'assigned value' 'and' unassigned value '):

$$\begin{cases} v(A \to B) = v(B) if\ v(A) = vd\ \ and\ v(B) = vnd, v(B) = vd\ \ and\ v(A) < v(B) \\ \qquad v(A \to B) = 1\ \ in\ the\ remaining\ cases \end{cases}$$

$$\begin{cases} v(A \wedge B) = vnd(\mathrm{v}(A), v(B))\ if\ and\ only\ \ if\ \ one\ of\ the\ values\ is\ not\ assigned \\ \qquad v(A \wedge B) = \max(v(A), v(B)) in\ the\ remaining\ cases \end{cases}$$

$$\begin{cases} v(A \vee B) = vd(v(A), v(B))\ when\ and\ only\ when\ one\ of\ \ the\ values\ is\ not\ assigned \\ \qquad v(A \vee B) = \min(v(A), v(B)) in\ the\ remaining\ cases \end{cases}$$

$$\{\ v(\neg A) = v(A) - 1$$

Theorem 4.5 Every C_ω-algebra can be extended to a C_ω-algebra with a least element.

Proof Let [S, ≡, →, ∧, ∨, '] be a C_ω-algebra that has no least element. Let 0 and 0 * be two elements not belonging to S. Let [S*, ≡*, →*, ∧*, ∨*, '*] be C_ω-algebra such that S* = S ∪ {0, 0*} and →*, ∧*, ∨*, '* operations such that they coincide, respectively, with →, ∧, ∨, ' in S. The extension ≡ (the fundamental relation of S), →, ∧, ∨, ' for S* is immediate:

(1) 0* ≡* 1 (where ≡* is the fundamental relation on S*).

(2) $0 <^* x, \forall x \in S \cup \{0^*\}$
(3) $0'^* = 0^*$
(4) $0'^{*'*} = 0'^* = 0$

$$(5)\begin{cases} 0 \to^* x = 1, x \in S^* \\ x \to^* 0 = 0, x \in S \cup \{0\} \\ x \to^* 0 = 1, x \in S^* \\ 0^* \to^* x = x, x \in S \cup \{0\} \end{cases}$$

S* is a C_ω-algebra with a least element, and S is a sub-C_ω-algebra of S*.

Definition 4.6 Let [S, ≡, →, ∧, ∨, '] be a C_ω-algebra. We call the implicative pre-lattice associated with S any isomorphic Curry system to the quotient algebra of [S, ≡, →, ∧, ∨,'] by the fundamental relation ≡.

Theorem 4.7 Every pre-lattice associated with [S, ≡, →, ∧, ∨, '] is a Heyting algebra or a maximal filter of a Heyting algebra (with or without the first element).

Proof If S has the first element, the Proof is immediate. If not, let S* be the C_ω-algebra obtained as in the preceding theorem. The quotient algebra S*/≡ is a Heyting algebra, and any algebra associated with S is isomorphic to the maximal filter of S*, which consists of all elements of this algebra except for its first element.

Theorem 4.8 Every C_ω-algebra is associated with a Heyting algebra of open sets of a topological space or with a maximal filter of such an algebra.

Proof Due to the previous theorem, all Heyting algebras are isomorphic to a Heyting algebra of open sets.

Theorem 4.9 Generally speaking, in a C_ω-algebra, no equivalence relations are non-trivial congruences.

2.5 Propositional C_n-Algebras

P.C. Rosenbloom's study of propositional Boolean algebra plays another interesting foundational role in the formalization of classical logic; see Rosenbloom [10]. In Boolean algebras, logical operations such as conjunction, disjunction, and negation are defined in algebraic terms, allowing for a structured analysis of classical propositional logic; see Rosenbloom [10]. Boolean algebras correspond to the logic of true and false values, where negation operates deterministically: the negation of a true proposition is false, and vice versa. While this framework provides a robust foundation for classical logic, it is limited when introducing contradictions or incomplete information.

In this context, propositional C_n-algebras emerge as a generalization of propositional Boolean algebra. The C_n-algebras, developed initially as algebraic structures corresponding to the C_n Calculus by da Costa, extend Boolean algebra to accommodate contradictions and indeterminacies that arise in paraconsistent and paracomplete logics; see da Costa [11]. In a propositional C_n-algebra, we redefine the classical operations, particularly negation, to handle propositions that might be both true and false or undecidable without leading to trivialization.

The move from C_n-algebras to propositional C_n-algebras involves a similar abstraction process as seen in Rosenbloom's work with Boolean algebras. Just as Boolean algebras are built upon classical logic's operations and equivalence relations, propositional C_n-algebras inherit their structure from C_n-algebras, but are explicitly designed to handle propositional formulas within the C_n Calculus. The key difference is the generalized treatment of negation and consistency, allowing for the coexistence of contradictory propositions and paracomplete (incomplete) information in a controlled manner; see da Costa [11].

It is important to note that while the theory of propositional C_n-algebras can be built directly from the broader theory of C_n-algebras, we focus on a specific subclass: propositional C_1-algebras, which are based on the C_1 Calculus—the simplest version of da Costa's paraconsistent systems. Propositional C_1-algebras serve as the foundational algebraic structures for reasoning in paraconsistent contexts.

Propositional C_1-algebras represent a concrete generalization of Boolean algebras that preserves many structural features but modifies key logical operations to account for paraconsistency. In particular, negation (complement) behavior in propositional C_1-algebras is non-classical. While Boolean algebras assume that a proposition and its negation cannot be true, propositional C_1-algebras allow for situations where both p and p' (complement or negation of p) hold simultaneously. This is accomplished by introducing a "consistency operator", similar to the one in C_n Calculus, which ensures that contradictions do not collapse the logical system into triviality; see da Costa [4, 11].

By focusing on propositional C_1-algebras, we restrict the general framework of propositional C_n-algebras to the simplest level of paraconsistency, which provides a manageable entry point for studying these systems algebraically. The reasoning within propositional C_1-algebras is structurally familiar to classical logicians—thanks to its connection to Boolean algebra—and strikingly novel due to the introduction of non-classical negation and the treatment of contradictions. This makes propositional C_1-algebras a crucial stepping stone toward the exploration of more complex propositional C_n-algebras.

It is interesting to note that propositional algebra also generalizes the concept of propositional algebra and Boolean algebra.

In [10], P. C. Rosenbloom developed the study known as propositional Boolean algebra. In this paragraph, we introduce the concept of propositional C_n-algebra, which generalizes the above concept. It is worth noting that if we start from C_n-algebras, $1 \leq n < \omega$, we can also build up the theory of propositional C_n-algebras, $1 \leq n < \omega$. However, as this last generalization is analogous to the first, we only deal with propositional C_1-algebras [12].

Definition 5.1 A propositional C_1-algebra is a Curry system [R, R*, $\equiv$, $\rightarrow$, $\wedge$, $\vee$, '], where $R^* \subseteq R$ (and $R \neq \emptyset$), $\rightarrow$, $\wedge$, $\vee$ are binary operations in R and $\neg$ is an operator in R, the following postulates are satisfied (where x^o abbreviates $\neg(x \wedge \neg x)$).

(A_1) $x, y \in R \Rightarrow x \to (y \to x) \in R^*$
(A_2) $x, y \in R \Rightarrow (x \to y) \to ((x \to (y \to z)) \to (x \to z)) \in R^*$
(A_3) $x, y \in R \Rightarrow x \in R^*$ and $x \to y \in R^* \Rightarrow y \in R^*$
(A_4) $x, y \in R \Rightarrow x \wedge y \to x \in R^*$
(A_5) $x, y \in R \Rightarrow x \wedge y \to y \in R^*$
(A_6) $x, y \in R \Rightarrow x \to (y \to (x \to y)) \in R^*$
(A_7) $x, y \in R \Rightarrow x \to (x \vee y) \in R^*$
(A_8) $x, y \in R \Rightarrow y \to (x \vee y) \in R^*$
(A_9) $x, y \in R \Rightarrow (x \to z) \to ((y \to z) \to (x \vee y \to z)) \in R^*$
(A_{10}) $x, y \in R \Rightarrow \neg\neg x \to x \in R^*$
(A_{11}) $x, y \in R \Rightarrow x \vee \neg x \in R^*$
(A_{12}) $x, y \in R \Rightarrow y \to ((x \to y) \to ((x \to \neg y) \to \neg x)) \in R^*$
(A_{13}) $x, y \in R \Rightarrow x^o \wedge y^o \to (x \to y)^o \in R^*$
(A_{14}) $x, y \in R \Rightarrow x^o \wedge y^o \to (x \wedge y)^o \in R^*$
(A_{15}) $x, y \in R \Rightarrow x^o \wedge y^o \to (x \vee y)^o \in R^*$
(A_{16}) $x, y \in R \Rightarrow x^o \to (\neg x)^o \in R^*$

Definition 5.2 Let R be a propositional C_1-algebra. Suppose that $x, y \in R$. So, $x \leftrightarrow y$ abbreviates $(x \to y) \wedge (y \to x)$. If $x \in R^*$, we also write $\vdash x$.

Theorem 5.3 All propositional Boolean algebras are a propositional C_1-algebra.

Proof Immediate.

Theorem 5.4 In a propositional C_1- algebra, we have:

(1) $\vdash x \to x$
(2) $\vdash x \wedge y \leftrightarrow y \wedge x$
(3) $\vdash (x \to y) \wedge (y \to x)$
(4) $\vdash x \vee y \leftrightarrow y \vee x$
(5) $\vdash (x \leftrightarrow y) \wedge (y \leftrightarrow z) \to (x \leftrightarrow z)$
(6) $\vdash x \wedge \neg(x \wedge \neg x) \to y$
(7) $\vdash y \to (x \to x)$
(8) $\vdash (x \wedge y) \wedge z \leftrightarrow x \wedge (y \wedge z)$
(9) $\vdash (x \to y) \to (x \wedge z \to y \wedge z)$
(10) $\vdash (x \vee y) \vee z \leftrightarrow x \vee (y \vee z)$
(11) $\vdash x \wedge (y \vee z) \leftrightarrow (x \wedge y) \vee (x \wedge z)$
(12) $\vdash x \vee (y \wedge z) \leftrightarrow (x \vee y) \wedge (x \vee z)$
(13) $\vdash x \vee (x \to y)$
(14) $\vdash ((x \to y) \to x) \to x$
(15) $\vdash (x')'$

Proof Immediate.

Definition 5.5 Let R be a propositional C_1-algebra and x, y elements of R.

Theorem 5.6 The relations $\equiv$ and $\leq$ defined above are, respectively, an equivalence relation and a quasi-order relation.

Proof We initially prove that $\equiv$ is an equivalence relation. By (1) of theorem 5.4 and by (A6), it follows that $\vdash x \leftrightarrow x$; therefore, $x \equiv x$. Suppose that $\vdash x \leftrightarrow y$, that is, that $\vdash x \rightarrow y \wedge y \rightarrow x$; by (3) of theorem 5.4, results in $\vdash y \rightarrow x \wedge x \rightarrow y$, where $\vdash y \leftrightarrow x$. So $x \equiv y$ *implies* $y \equiv x$. If $x \equiv y$ and $y \equiv z$, $x \equiv z$ results, applying (5) of theorem 5.4 and (A6).

We now prove that $\leq$ is a quasi-order. As $\vdash x \rightarrow x$ it is seen that $x \leq x$. Also, from $\vdash x \rightarrow y$ and $\vdash y \rightarrow x$, we obtain $\vdash x \rightarrow z$, resulting in $x \leq z$.

Theorem 5.7 Let R be a propositional C_1-algebra. We have to:

(1) $x \leq y$ if, and only if $x \wedge y \equiv x$
(2) $x \leq y$ if, and only if $x \vee y \equiv y$
(3) $\forall x \in \mathrm{R}, 0 \leq x$ *and* $x \leq 1$.

Proof The Proof of (1) and (2) is immediate. (3) is a corollary of (6) and (7) of Theorem 5.4.

Theorem 5.8 If [R, R*, $\equiv$, $\rightarrow$, $\wedge$, $\vee$, $\neg$] is a propositional C_1-algebra, if we define $=$ and 0 (a least element) as previously and ' abbreviates $\neg$, then [R, R*, $\equiv$, $\rightarrow$, $\wedge$, $\vee$, 0, '] is a C_1-algebra.

Proof Immediate.

Theorem 5.7 If [R, $\equiv$, $\rightarrow$, $\wedge$, $\vee$, 0, '] forms a C_1-algebra, defining R* as the set of elements $x \in \mathrm{R}$ such that $x \equiv y \rightarrow y$, for some $y \in \mathrm{R}$, and representing the operator ' by $\neg$, [R, R*, $\equiv$, $\rightarrow$, $\wedge$, $\vee$, $\neg$] is a propositional C_1-algebra.

Proof Immediate.

The previous propositions show that C_1-algebra and C_1-propositional algebra are equivalent structures (in a sense that generalizes Bourbaki). In this way, everything that is valid for one of them can be adapted to the other. For example, for propositional C_1-algebras, representation theorems such as Theorems 2.17 and 2.18 are proved.

Definition 5.10 In a C_1-propositional algebra, [R, R*, $\equiv$, $\rightarrow$, $\wedge$, $\vee$, $\neg$], the operator $\neg$ is not monotonous concerning $\equiv$ (fundamental relation, definition 5.5).

Proof Similar to Theorem 2.7.

Definition 5.11 Let R be a propositional C_1-algebra. We define:
$\mathrm{S}^{o} = \{x \mid x \in \mathrm{R} \text{ and } \vdash x^{o}\}$
$(\mathrm{S}^{o})^{*} = \{x \mid x \in \mathrm{R} \text{ and } \vdash x^{o} \wedge x\}$.

Theorem 5.12 Let [R, R*, $\equiv$, $\rightarrow$, $\wedge$, $\vee$, $\neg$] be a propositional C_1-algebra. So, [R°, (R°)*, $\equiv$, $\rightarrow$, $\wedge$, $\vee$, $\neg$] constitutes a propositional Boolean algebra.

Proof It must be shown that the postulates that define the concept of propositional Boolean algebra, according to Rosenbloom [10], p. 31, are true. The proof, although laborious, offers no essential difficulties.

Theorem 5.13 Consider C_1-algebra R, ≡, →, ∧, ∨, ¬], and let $\{x \mid x \in R \text{ and } x \equiv 1\}$; in these conditions, [R°, ∧, ∨, ’] is a Boolean algebra.

Proof Immediate.

Intuitively, Theorems 5.12 and 5.13 show that, in C_1-algebras or C_1-propositional algebras, the “well-behaved” elements, i.e., those that satisfy the Principle of Non-contradiction, behave in a “classic” way.

Definition 5.14 In a propositional C_1-algebra, as in C_1-algebras, x^* abbreviates $x^o \wedge x'$.

Theorem 5.15 If [R, ≡, →, ∧, ∨, ¬] is a propositional C_1-algebra, then [R, ≡, →, ∧, ∨, *] is an algebra propositional Boolean, with the operator *.

Proof A version of Theorem 2.14 for propositional C_1-algebras.

2.6 Deductive Systems

Tarski and Stone developed the theory of deductive systems. This section shows how such a theory can be extended to C_1-algebras. We will study the question only in C_1-algebras and C_1-propositional algebras; the generalization of the results for C_n-algebras, $1 \leq n < \omega$, and for propositional C_n-algebras, $1 \leq n < \omega$, is immediate [5].

Definition 6.1 Consider a C_1-algebra. If x is an element of this algebra, $\vdash x$ abbreviates $x \equiv 1$.

The exposition shows that the symbols “ $\vdash$”, “≡”, “ ≤ ”, etc., can be used, in a similar way, both in C_1-algebras and in C_1-propositional algebras. Thus, the two structures are equivalent.

Definition 6.2 In a C_1-algebra, we call a filter a non-empty subset F of R meeting the following conditions:

(1) For all $x, y \in F \Rightarrow x \wedge y \in F$
(2) For all $x, \in F \Rightarrow x \vee y \in F$
(3) For all $x, x \rightarrow y \in F \Rightarrow y \in F$
(4) If $x \in F$ and $x \equiv y \Rightarrow y \in F$

Theorem 6.3 A necessary and sufficient condition for the subset F of a C_1-algebra to be a filter is to have:

(1) $1 \in F$
(2) For all x, $x \to y \in F \Rightarrow y \in F$
(3) If $x \in F$ and $x \equiv y \Rightarrow y \in F$

Proof ($\Leftarrow$). As $1 \in F$, $F \neq \emptyset$. If $x, y \in F$, as $x \to (y \to x \wedge y) \equiv 1 \in F$ results in $x \wedge y \in F$. Since $x \to x \vee y \equiv 1 \in F$, if $x \in F$, then $x \vee y \in F$, whatever the element y of that algebra.

($\Rightarrow$). Suppose that F is a filter. So, $x \vee y \in F$, Therefore, there is $x \in F$. Consequently, $x \vee 1 \equiv 1 \in F$.

Theorem 6.4 If F is a filter, $x \in F$ and $x \leq y$ yield $y \in F$.

Proof $x \in F$ and $x \leq y$ result in $x \to y \equiv 1$. Thus, $x \in F$ and $x \to y \in F$, and therefore $y \in F$.

Theorem 6.5 For a subset F of R to be the filter of a C_1-algebra $[R, \equiv, \to, \wedge, \vee, \neg]$ it is necessary and sufficient that one of the conditions I or II is verified:

(I)

a) $F \neq \emptyset$
b) $x, y \in F \Rightarrow x \wedge y \in F$
c) $x \in F$ and $x \leq y \Rightarrow y \in F$

(II)

a. $F \neq \emptyset$
b. $x \wedge y \in F \Leftrightarrow x \in F$ and $y \in F$
c. $x \in F$ and $x \equiv y \Rightarrow y \in F$

Proof Immediate.

Theorem 6.6 A necessary and sufficient condition for the subset F of a C_1-algebra $[R, \equiv, \to, \wedge, \vee, \neg]$ to be a filter, is that, whatever the elements x, y, and z belonging to that algebra, such that:

(F_1) $x, y \in R \Rightarrow x \to (y \to x) \in F$
(F_2) $x, y, z \in R \Rightarrow (x \to y) \to ((x \to (y \to z)) \to (x \to z)) \in F$
(F_3) $x, y \in R \Rightarrow x \in F$ e $x \to y \in F \Rightarrow y \in F$.
(F_4) $x, y \in R \Rightarrow x \wedge y \to x \in F$.
(F_5) $x, y \in R \Rightarrow x \wedge y \to y \in F$.
(F_6) $x, y \in R \Rightarrow x \to (y \to (x \to y)) \in F$.
(F_7) $x, y \in R \Rightarrow x \to (x \vee y) \in F$.
(F_8) $x, y \in R \Rightarrow y \to (x \vee y) \in F$.
(F_9) $x, y, z \in R \Rightarrow (x \to z) \to ((y \to z) \to (x \vee y \to z)) \in F$.
(F_{10}) $x, y \in R \Rightarrow \neg\neg x \to x \in F$.
(F_{11}) $x, y \in R \Rightarrow x \vee \neg x \in F$.
(F_{12}) $x, y \in R \Rightarrow y^{o} \to ((x \to y) \to ((x \to \neg y) \to \neg x)) \in F$.
(F_{13}) $x, y \in R \Rightarrow x^{o} \wedge y^{o} \to (x \to y)^{o} \in F$.
(F_{14}) $x, y \in R \Rightarrow x^{o} \wedge y^{o} \to (x \wedge y)^{o} \in F$.

(F_{15}) $x, y \in R \Rightarrow x^o \wedge y^o \rightarrow (x \vee y)^o \in F$.
(F_{16}) $x, y \in R \Rightarrow x^o \rightarrow (\neg x)^o \in F$.
(F_{17}) $x, y \in R \Rightarrow x \in F$ and $x \equiv y \Rightarrow y \in F$.

Proof An immediate consequence of the previous theorem.

Theorem 6.7 A filter A in a C_1-algebra is a filter in A considered as a Boolean algebra, and vice versa.

Proof Immediate.

Theorem 6.8 Let [R, R*, ≡, →, ∧, ∨, ¬] be a propositional C_1-algebra. So, R* is a filter.

Proof An immediate consequence of the definition of a propositional C_1-algebra.

Definition 6.9 Let [R, ≡, →, ∧, ∨, '] be a C_1-algebra. Let M be a subset of R, the finite sequence $x_1, x_2, \ldots, x_n$ of elements of R is called a deduction of x (where x is an element of R) from the hypotheses M if:

(1) For all i, $1 \leq i \leq n$, or $x_i \in M$ or $\vdash x_i$ or there are j, k < i such that x_k is $x_j \rightarrow x_i$;
(2) x_n is x;

If there is a deduction of x from hypotheses M, x is said to be a consequence of M and symbolized by M $\vdash x$.

If M $\neq \emptyset$, the deduction is called a (formal) proof, and the sign "$\vdash$" regains its original meaning, writing $\vdash x$ instead of $\emptyset \vdash x$. (See theorem 6.11 below).

When we have $\vdash x$, x is said to be provable. The provable elements, therefore, form a filter.

The set of consequences of M is represented by M #.

If M = $\{x_1, x_2, \ldots, x_n\} \vdash x$, we write $x_1, x_2, \ldots, x_n \vdash x$.

Theorem 6.10 For any elements $x_1, x_2, \ldots, x_n, x$ of a C_1-algebra, one has $x_1, x_2, \ldots, x_n \vdash x$ if and only if $x_1 \wedge x_2 \wedge \ldots \wedge x_n \vdash x$.

Proof A consequence of the fact that all C_1-algebras are an implicative lattice.

Theorem 6.11 Consider a C_1-algebra. We then have:

(1) $\emptyset \vdash x$ if and only if $\vdash x$.
(2) If M $\neq \emptyset$, M $\vdash x$ if and only if there are elements $x_1, x_2, \ldots, x_n \in M$ such that que $x_1, x_2, \ldots, x_n \vdash x$.
(3) $x \vdash y$ if and only if $\vdash x \rightarrow y$

Proof Immediate.

Definition 6.12 In a C_1-algebra, the filter of the demonstrable elements will be denoted by D.

Theorem 6.13 In a C_1-algebra, M# is the smallest filter containing D and M.

Proof Immediate.

Theorem 6.14 In propositional C_1-algebra [R, R*, $\equiv$, $\rightarrow$, $\wedge$, $\vee$, $\neg$], M ## = M #; $\{x \wedge x' \wedge x^{o}\}\# = \{x, x', x^{o}\}\# = R$; $x \rightarrow y \in M\# \Leftrightarrow y \in (M \cup \{x\})\#$; $\emptyset^* = \{1\}\#\mu = R^*$ ($M \subseteq R$ and x and y are assumed to be any elements of R).

Definition 6.15 A subset M of a C_1-algebra is said to be a deductive system or a theory if M # $\subseteq$ M.

Theorem 6.16 Let R be a C_1-algebra. We have:

(1) If M e is a deductive system, then M # = M.
(2) $\{x\}\# = \{x^*\}\#$

Proof Immediate.

Definition 6.7 Consider a C_1-algebra R. A subset M of R is said to be trivial if M $\vdash 0$.

Definition 6.18 Under the conditions of the previous definition, M is called inconsistent if M $\vdash x$ and M $\vdash x'$, for some x belonging to R. Otherwise, M is said to be consistent.

Theorem 6.19 Let R be a C_1-algebra. If F is a maximal filter, for every element x of R, $x \in$ R, or $x^* \in$ R.

Proof Immediate.

Definition 6.20 The deductive system M is said to be complete if it is a maximal filter.

Theorem 6.21 The deductive system M of C_1-algebra R is complete if and only if for any element x of R, $x \in$ M, or $x^* \in$ M.

Proof Immediate.

Theorem 6.22 Let M be a deductive system in C_1-algebra R. Let us consider.
RM = $\{x \mid x \in R \text{ and } \vdash x\}$ and
$M^{o} = \{x \mid x \in M \cap RM\}$
So, M^{o} is a deductive system in Boolean RM algebra.

Theorem 6.23 There are inconsistent and non-trivial deductive systems.

Proof Let R be C_1-free algebra "generated by two generators" x and y. In R, the relation $x \wedge x' \leq y$ or what is equivalent to $x \wedge x' \rightarrow y \equiv 1$, is not valid. In effect, just apply the matrix [{1, 2, 3}, {1, 2}, $\rightarrow$, $\wedge$, $\vee$, '], where the operations $\rightarrow$, $\wedge$, $\vee$, ' are defined by the following Table 2.1.

We consider the deductive system M = $\{x, x'\}\#$ in R. M is, of course, inconsistent but not trivial, given that y [R, R*, $\equiv$, $\rightarrow$, $\wedge$, $\vee$, $\neg$], M, because, otherwise, we would have that $x \wedge x' \rightarrow y \equiv 1$.

There are other relevant examples of inconsistent and non-trivial systems. The reader can consult, for example, [13] and [14].

Table 2.1 Operations →, ∧, ∨, ’

x	y	$x \to y$	$x \wedge y$	$x \vee y$	x’
1	1	1	1	1	3
2	1	1	1	1	1
3	1	1	3	1	1
1	2	1	1	1	
2	2	1	1	1	
3	2	1	3	1	
1	3	3	3	1	
2	3	3	3	1	
3	3	1	3	3	

2.7 Sub-Algebras and Homomorphisms

Definition 7.1 Let [R, ≡, →, ∧, ∨, ¬] be a C_1-algebra. A sub-algebra of R is a subset B of R such that B, with the fundamental relation ≡ of R, the distinguished elements and operations of R, is a C_1-algebra.

Every C_1-algebra contains a trivial sub-algebra, namely, 2; all other sub-algebras will be called non-trivial. Every C_1-algebra contains an improper sub-algebra, namely, R; all other sub-algebras will be called their own.

Theorem 7.2 The intersection of each collection of sub-algebras of a C_1-algebra R is a sub-algebra of R.

Proof Immediate.

Definition 7.3 The theorem above allows us to define the concept of sub-algebra generated by a subset of R: if E is any subset of R, then the intersection of all sub-algebras of R that contains E, is an E# sub-algebra. Such a sub-algebra is called the sub-algebra generated by E.

Definition 7.4 Let [R, ≡, →, ∧, ∨, ’] and [R*, ≡*, →*, ∧*, ∨*, ’*] two C_1-algebras. A homomorphism of R to R* is a function f of R to R* such that it preserves the operations and the distinguished elements, i.e.,

(H_1) f($x \to y$) ≡* f(x) →* f(y)
(H_2) f($x \wedge y$) ≡* f(x) ∧* f(y)
(H_3) f($x \vee y$) ≡* f(x) ∨* f(y)
(H_4) f(x’) ≡* f(x)’*
(H_5) if $x \equiv y$, then, f(x) ≡* f(y)

Some particular types of homomorphisms can be described, as in classical algebraic theory. A monomorphism is a homomorphism such that if f (x) ≡* f (y), then $x \equiv y$. An epimorphism is a homomorphism such that if $y \in$ R*, there is $x \in$ R such that $x \equiv y$. An isomorphism is a homomorphism that is both a monomorphism and an epimorphism. If R* ⊆ R, f is called an endomorphism; if it is an isomorphism, then f is called an automorphism. If there is an isomorphism of R and R*, then R and R* are said to be isomorphic.

Theorem 7.5 If f of R in R* is a homomorphism of C_1-algebras, then the homomorphic image f(R) is a sub-algebra of R*.

Proof Immediate.

Definition 7.6 Let f be a homomorphism of C_1-algebras of R in R*. The core (or kernel) of f is the set of elements of R whose image by f is associated with element 0 of R*, that is,

$\text{Ker(f)} = \{x \in \text{R} \mid \text{f}(x) \equiv^* 0^*\}$

The co-nucleus (or shell) of f is the set of elements of R whose image by f is associated with element 1 of R*, that is.

$\text{Sh(f)} = \{x \in \text{R} \mid \text{f}(x) \equiv^* 1^*\}$

Theorem 7.7 The core of any homomorphism is an ideal. Also, the co-nucleus of a homomorphism is a filter.

Proof Immediate.

Theorem 7.8 A necessary and sufficient condition for a homomorphism to be a monomorphism is that.

$\text{Ker(f)} = \{x \in \text{R} \mid \text{f}(x) \equiv^* 0^*\} = \{0\}$

(the class of R elements associated with 0).

Proof Immediate.

2.8 Soundness and Completeness of C_1

In this paragraph, we apply the notions of homomorphism and ultrafilter to demonstrate the Soundness and completeness theorems of the C_1 calculus. Let us consider, then, the C_1-algebra [F, $\equiv$, $\rightarrow$, $\wedge$, $\vee$, ’] obtained in theorem 2.6, Chap. II.

Theorem 8.1 If the shell of a homomorphism f of F in any C_1-algebra is an ultrafilter, then:

(1) if $\text{f}(x) \equiv 0$, then $\text{f}(x') \equiv 1$.
(2) if $\text{f}(x'') \equiv 1$, then $\text{f}(x) \equiv 1$.
(3) if $\text{f}(x') \equiv \text{f}(x \rightarrow y) \equiv \text{f}(x \rightarrow y') \equiv 0$, then $\text{f}(x) \equiv 1$.
(4) $\text{f}(x \rightarrow y) \equiv 1$ if and only if $\text{f}(x) \equiv 0$ ou $\text{f}(y) \equiv 1$.
(5) $\text{f}(x \wedge y) \equiv 1$ if and only if $\text{f}(x) \equiv 1 \equiv \text{f}(y) \equiv 1$.
(6) $\text{f}(x \vee y) \equiv 1$ if and only if $\text{f}(x) \equiv 1$ ou $\text{f}(y) \equiv 1$.
(7) if $\text{f}(x^o) \equiv \text{f}(y^o) \equiv 1$, then $\text{f}((x \rightarrow y)^o) \equiv \text{f}((x \wedge y)^o) \equiv \text{f}((x \vee y)^o) \equiv 1$.

Proof Immediate.

Definition 8.2 Let F be the class of C_1 formulas, and f be a homomorphism of F in a C_1-algebra. We write f $\models \Gamma$, where Γ is a subset of F, if for each $x \in \Gamma$, f(x) $\equiv 1$. $\Gamma \models x$ means that for all homomorphisms f of F in an arbitrary C_1-algebra if f $\vdash \Gamma$, then f (x) $\equiv 1$.

Theorem 8.3 (Soundness): If x is a provable formula of C_1, then $\models x$.

Proof By induction on the length of the tests.

Theorem 8.4 Let U be an ultrafilter in F. So, there is a homomorphism f of F into 2 such that the shell of f is exactly U.

Proof Immediate.

Theorem 8.5 (Completeness). Let F be the class of the formulas of C_1, and $x \in$ F. Suppose that f(x) $\equiv 1$ for any homomorphism f of F in an arbitrary C_1-algebra. So, x is a provable formula for C_1. [7, 15]

Proof Immediate.

References

1. da Costa, N.C.A.: Opérations non monotones dans les treillis. C.R. Acad. Sc. Paris 263, 429–432 (1966)
2. da Costa, N.C.A.: Álgebras de Curry, (in Portuguese, mimeographed) University of São Paulo, São Paulo, (1967)
3. da Costa, N.C.A.: Filters et idéaux d'une algebrie C_n, C.R. Acad. Sc. Paris 264, 549–552 (1967)
4. da Costa, N.C.A.: On the theory of inconsistent formal systems', Notre Dame. J. of Formal Logic **15**, 497–510 (1974). https://doi.org/10.1305/ndjfl/1093891487
5. Barros, C.M., N.C.A. Da Costa, Abe, J.M.: Tópico de Teoria dos Sistemas Ordenados, Vol II: Sistemas de Curry, Coleção Documentos, Série Lógica e Teoria da Ciência, IEA-USP, 20, p. 132, (1995). ISSN 16799429
6. Sikorski, R.: Boolean Algebras, Springer-Verlag, (1969). ISBN-13: 978-3540044697, ISBN-10: 3540044698
7. Mortensen, C.: Every quotient algebra for C1_is trivial. Notre Dame J. Formal Log. **21**(4) (1980). https://doi.org/10.1305/ndjfl/1093883254
8. Henkin, L., Monk J.D., Tarski, A.: Cylindric Algebras, Part I, lecture notes in mathematics (Book 883), pages 323, (1981). ISBN-13: 978–3540108818; ISBN-10: 3540108815
9. Curry, H.B.: Foundations of mathematical logic, dover books on mathematics 2nd ed., 416 pages 2010 ISBN-10: 0486634620; ISBN-13: 978–0486634623
10. Rosenbloom, P.C.: The elements of mathematical logic, Dover, (2005). ISBN-10: 0486446174, ISBN-13: 978–0486446172
11. da Costa, Newton C.A., Marconi, Diego.: A note on paracomplete logic. Atti della Accademia Nazionale dei Lincei. Classe di Scienze Fisiche, Matematiche e Naturali. Rendiconti Lincei. Matematica e Applicazioni 80.7–12, 504–509. (1986). <http://eudml.org/doc/287306>
12. Abe, J.M.: Curry algebras and propositional algebra C_1. Int. J. Knowl.-Based Intell. Eng. Syst., ISSN: 1327–2314 (Print), **5**(2), 127–132, (2013)

13. Abe, J.M., Akama, S., Nakamatsu.: Introduction to Annotated Logics—Foundations for Paracomplete and Paraconsistent Reasoning, Series Title Intelligent Systems Reference Library, Volume 88, Publisher Springer International Publishing, Copyright Holder Springer International Publishing Switzerland, eBook ISBN 978–3–319–17912–4, https://doi.org/10.1007/978-3-319-17912-4, Hardcover ISBN 978–3–319–17911–7, Series ISSN 1868–4394, Edition Number 1, 190 pages, (2015)
14. Priest, G.: An Introduction to Non-Classical Logic: From If to Is. Cambridge University Press (2002)
15. Bourbaki, N.: Elements of mathematics: Set theory. Springer (2004)

Chapter 3
Heyting Algebras

Heyting algebras are algebraic structures that provide a formal framework for intuitionistic logic, a system of logic that rejects the Principle of Excluded Middle; see Heyting [1, 2]. Developed by A. Heyting in the context of constructivist mathematics, Heyting algebras generalize Boolean algebras by allowing for a more nuanced treatment of truth and implication, making them particularly suitable for reasoning about constructive proofs, where a statement is only considered true if there is an explicit construction or proof for it; see Troelstra & van Dalen [3].

Unlike Boolean algebras, in which every proposition must be either true or false, Heyting algebras accommodate the possibility that a proposition may neither be proven true nor proven false, reflecting the constructive nature of the logic. This flexibility has led to the application of Heyting algebras not only in logic and mathematics but also in fields like computer science, particularly in the study of type theory and constructive proofs, as well as in topology, where they offer a natural framework for describing open sets and continuous functions; see Johnstone [4].

This chapter explores several important themes related to Heyting algebras, including constructivity, ideality, and intentional operators. Constructivity, a central concept to intuitionistic logic, refers to the idea that mathematical objects must be constructed explicitly rather than assumed to exist ideally. This constructivist perspective is reflected in Heyting algebras through the absence of the Principle of Excluded Middle, which prevents the assumption of a proposition's truth or falsehood without proof; see Dummett [5].

In the context of Heyting algebras, Ideality often refers to certain subsets or filters within the algebra that preserve certain logical properties analogous to ideals in ring theory. These ideals can be used to study the behavior of elements within the algebra and explore relationships between propositions; see Bezhanishvili & Gehrke [6].

Intentional operators are algebraic operations that capture various aspects of reasoning within Heyting algebras, particularly those related to the intentionality of mathematical constructions. These operators help to formalize how propositions are constructed, related, and evaluated within a constructive framework.

J. M. Abe, *An Introduction to Curry Systems*, Intelligent Systems Reference Library 280, https://doi.org/10.1007/978-3-031-97810-4_3

In addition to these themes, we will explore how the structures of Heyting algebras can be generalized and extended to address concepts beyond logic, particularly in computability and topological spaces. Establishing a link between Heyting algebras and topological spaces makes it possible to define open sets and continuous mappings within a constructive framework, offering new perspectives on the interplay between logic, topology, and computation; see Goldblatt [7].

This chapter will give an initial idea via Curry systems of how these notions can be defined and extended, drawing connections between Heyting algebras and other mathematics and computer science areas. In doing so, we aim to show how Heyting algebras provide a flexible and powerful tool for formal reasoning, particularly in settings where constructivity and computability are essential.

3.1 Heyting Algebras and Heyting Propositional Algebras

Definition 1.1 We will call a Heyting algebra any absolute implicative pre-lattice with the first element [H, $\equiv$, $\rightarrow$, $\wedge$, $\vee$, 0]. As implicative pre-lattices, Heyting algebras have a greater element.

Definition 1.2 In a Heyting algebra, $\neg x$ shortens $x \rightarrow 0$ and $\vdash x$ shortens $x \equiv 1$.

Example 1.3 Let S be a topological space, and we designate "o" as the interior operator. If A and B are subsets of S, $A \rightarrow B$ will denote the set $((S - A) \cup B)^{o}$ and $\neg A$ denotes $A \rightarrow \emptyset$.

Let S* be a subset of P (S) (set of parts of S), such that if A, B, $\in$ S* , $A \cup B$, $A \cap B$ and $\neg A$ also belong to S* . So [S * , $\equiv$, $\rightarrow$, $\wedge$, $\vee$, $\emptyset$] (where the fundamental relation $\equiv$ is equality $=$) is a Heyting algebra and is called an open algebra.

Theorem 1.4 Any Heyting algebra, [H, $\equiv$, $\rightarrow$, $\wedge$, $\vee$, 0] is such that H/$\equiv$ is isomorphic to an open algebra.

Proof We leave it as an exercise.

Definition 1.5 A structure [H, H*, $\equiv$, $\rightarrow$, $\wedge$, $\vee$, 0] is called Heyting's propositional algebra if H $\neq \emptyset$, H* $\subseteq$ H, $\rightarrow$, $\wedge$ and $\vee$ are binary operations on H and $\neg$ operator in H), satisfying the postulates:

(H_1) $x, y \in H \Rightarrow x \rightarrow (y \rightarrow x) \in H^*$
(H_2) $x, y, z \in H \Rightarrow (x \rightarrow y) \rightarrow ((x \rightarrow (y \rightarrow z)) \rightarrow (x \rightarrow z)) \in H^*$
(H_3) $x, y, \in H \Rightarrow x \in H^*$ and $x \rightarrow y \in H^* \Rightarrow y \in H^*$
(H_4) $x, y, \in H \Rightarrow x \wedge y \rightarrow x \in H^*$
(H_5) $x, y, \in H \Rightarrow x \wedge y \rightarrow x \in H^*$
(H_6) $x, y, \in H \Rightarrow x \rightarrow y \rightarrow (x \rightarrow y)) \in H^*$
(H_7) $x, y, \in H \Rightarrow x \rightarrow (x \vee y) \in H^*$
(H_8) $x, y, \in H \Rightarrow y \rightarrow (x \vee y) \in H^*$
(H_9) $x, y, \in H \Rightarrow (x \rightarrow z) \rightarrow ((y \rightarrow z) \rightarrow (x \vee y \rightarrow z)) \in H^*$

(H_{10}) $x, y, \in H \Rightarrow (x \to y) \to ((x \to \neg y) \to \neg x) \in H^*$
(H_{11}) $x, y, \in H \Rightarrow \neg x \to (x \to y) \in H^*$

Definition 1.6 Let H be a propositional Heyting algebra. So:

$\vdash x$ abbreviates $x \in H^*$
0 abbreviates $x \wedge \neg x$.
$x \leq y$ abbreviates $x \to y \in H^*$
$x \equiv y$ abbreviates $(x \to y) \wedge (y \to x) \in H^*$

Theorem 1.7 Let $[H, \equiv, \to, \wedge, \vee, 0]$ be a Heyting algebra and put $H^* = \{x \mid x \in H \text{ and } x \equiv 1\}$. So, $[H, H^*, \equiv, \to, \wedge, \vee, \neg]$ ($\neg$ defined according to definition 1.2), is a propositional Heyting algebra.

Proof Immediate.

Theorem 1.8 If $[H, H^*, \equiv, \to, \wedge, \vee, \neg]$ is a propositional algebra from Heyting, $[H, \equiv, \to, \wedge, \vee, 0]$ and "0" is defined as in Definition 1.6, then it is a Heyting algebra.

Proof Immediate.

Theorems 1.7 and 1.8 show that the concepts of Heyting algebra and propositional Heyting algebra are equivalent. For such structures, the concepts of filter, deductive system, etc., can be defined, as was done for C_1-algebras and C_1-propositional algebras.

Heyting's algebras and Heyting's propositional algebras are algebraic versions of intuitionist propositional calculus.

3.2 The Constructibility Operator

We now develop a classic algebraic theory of intuitionist propositional calculus. For that, we start from a pre-Boolean algebra of $[B, \equiv, \wedge, \vee, ', 0, 1, ^\circ]$, in which the operator "$^\circ$" is defined.

The intuitive interpretation of this operator is as follows: if x represents a proposition, x° means that proposition x is intuitionistically true (or constructively true). In other words, x expresses that there is an appropriate construction for x; that is, that x is true by construction; see [8].

The "$^\circ$" operator must, of course, satisfy certain postulates, which we enunciate below.

Definition 2.1 Let $[B, \equiv, \wedge, \vee, ', 0, 1, ^\circ]$ be a Boolean pre-algebra with the operator "$^\circ$". If $x, y \in B$, then $x \to^\circ y$ abbreviates $(x \to y)^\circ$ e $\neg^\circ x$ abbreviates $x \to^\circ 0^\circ$. Also, we denote by Bo the image of B by "$^\circ$".

That said, the following are postulated by the operator "$^\circ$".

(C_1) $x^{oo} \equiv x^o$
(C_2) $x^o \equiv\leq x$
(C_3) $(x \wedge y)^o \equiv x^o \wedge y^o$

(C_4) if $x, y \in B^o$, then $x \wedge y, x \vee y \in B^o$
(C_5) if $x, y, z \in B^o$, then $(x \wedge z \leq y \Rightarrow z \leq x \rightarrow^o y)$

The intuitive meanings of the above postulates are apparent. For example, (C_1) states that it is constructively true that x is constructively true, which is equivalent to saying that x is constructively true. (C_2) means that if x is constructively true, then x is classically true.

Theorem 2.2 Let $[B, \equiv, \wedge, \vee, ', 0, 1, {}^\circ]$ be a Boolean algebra with the operator "o". We then have:

(1) $0^\circ \equiv 0$
(2) if $x \in B$, then $x'^o \leq x'$
(3) if $x \in B$, then $x' \leq x^{o\prime}$
(4) For all $x, y \Rightarrow x \rightarrow^o y \leq x \rightarrow y$
(5) For all $x, y \Rightarrow x \wedge (x \rightarrow^o y) \leq y$

Proof

(1) $0 \leq 0^\circ$, since 0 is the first element of Boolean's pre-algebra under consideration. But, by (C_2), $0^\circ \leq 0$. Therefore, $0^\circ \equiv 0$.
(2) A consequence of (C_2).
(3) Immediate.
(4) $x \rightarrow^o y$ is $(x \rightarrow y)^o$ However, $(x \rightarrow y)^o \leq x \rightarrow y$. Therefore, $x \rightarrow^o y \leq x \rightarrow y$.
(5) In a Boolean pre-algebra, $x \wedge (x \rightarrow y) \leq y$, since $x \rightarrow y$ abbreviates $x' \vee y$. Therefore, due to (4), $x \wedge (x \rightarrow^o y) \leq y$.

Therefore, under (4), $x \wedge (x \rightarrow^o y) \leq y$.

Definition 2.3 If x is an element of Boolean algebra $[B, \equiv, \wedge, \vee, ', 0, 1, {}^\circ]$, 1_o abbreviates $x^o \rightarrow^o x^o$.

Theorem 2.4 With the previous notations $[B, \equiv, \rightarrow^\circ, \wedge, \vee, \neg^\circ, 0^\circ, 1_\circ]$ is a Heyting algebra.

Proof Heyting's algebras are absolute implicative pre-lattices with the first element.

We initially prove that $[B^o, \equiv, \rightarrow^o, \wedge, \vee]$ is a pre-lattice. In fact, by (C_4), if $x, y \in B^o$, then $x \wedge y, x \vee y \in B^o$. Naturally, operations $\wedge$ and $\vee$ satisfy, since any Boolean pre-algebra is a pre-lattice, the postulates of this last type of structure.

For $[B^o, \equiv, \rightarrow^o, \wedge, \vee]$ to be an implicative pre-lattice, it is necessary to have:

(1) For all $x, y \in B^o \Rightarrow x \wedge (x \rightarrow^o y) \leq y$
(2) If $x, y, z \in B^o$, then $(x \wedge z \leq y \Rightarrow z \leq x \rightarrow^o y)$.

By Theorem 2.2, part (4), (1) is true; from the postulate (C_5), it appears that (2) also applies. Therefore, [B^o, ≡, $\rightarrow^o$, ∧, ∨] is pre-lattice implicit. As $0^o \equiv 0$, 0° is the first element of the pre-lattice in question. By definitions 2.1 and 2.3, $\neg^o$ and 1° are, respectively, the negation operator and the last element of such a pre-lattice.

It will be proved below that [B^o, ≡, $\rightarrow^o$, ∧, ∨, $\neg^o$, 0°, 1°] is pre-lattice absolute implicative (Peirce's law does not apply).

Theorem 2.5 Let [B^o, ≡, $\rightarrow^o$, ∧, ∨, $\neg^o$, 0°, 1°] be the pre-lattice in question in the previous theorem. We have:

(1) $1^o \leq 1_o$
(2) If $x, y \in^o B^o$, then $x \equiv y \rightarrow x^o \equiv y^o$.

Proof (2) If $x, y \in B^o$, then there are $x_1, y_1 \in B^o$, then such that $x \equiv x_1{}^o$ e $y \equiv y_1{}^o$. So, $x^o \equiv x_1{}^{oo}$ and $y^o \equiv y_1{}^{oo}$. As $x \equiv y$, i.e., $x_1{}^o \equiv y_1{}^o$, it follows that $x^o \equiv y^o$.

Theorem 2.6 Let [B^o, ≡, $\rightarrow^o$, ∧, ∨, $\neg^o$, 0°, 1°] be a Boolean pre-algebra with the operator "o". If $x, y \in B^o$, $x \equiv y$ does not imply that $x^o \equiv y^o$

Proof Just consider Boolean's pre-algebra

$$[\{0, a, 1\}, \wedge, \vee, ', 0, 1, {}^o]$$

with the operator "o", where the basic equivalence relation ≡ is defined by {(0, 0), (a, a), (a, 1), (1, 0), (1, 1)}. Operations ∧, ∨, ', ^{o} are characterized by Table 3.1.

x	x'	x	x^o
0	1	0	0
a	0	a	0
1	0	1	1

It appears that all the postulates are satisfied. However, $x \equiv y$ does not entail $x^o \equiv y^o$; in fact, $a \equiv 1$, but we do not have $a^o \equiv 1^0$ (where $a^o = 0$ and $1^0 = 1$).

The two preceding theorems show that the operator "o" is not monotonous concerning ≡ in [B, ≡, ∧, ∨, ', 0, 1, °], but is monotonous in [B^o, ≡, $\rightarrow^o$, ∧, ∨, $\neg^o$, 0°, 1_o]. It seems reasonable since, for example, in classical predicate calculus, one has

Table 3.1 Operations ∧, ∨, ', ^{o}

$x \wedge y$	$x \backslash y$	0	a	1	$x \vee y$	$x \backslash y$	0	a	1
	0	0	0	0		0	0	a	1
	a	0	a	a		0	a	a	1
	1	0	a	1		1	0	1	1

(1) $\neg\forall x[\neg A(x) \leftrightarrow \exists x A(x)]$, however, from (1), it cannot be deduced that $\vdash$ $(\neg\forall x\neg A(x))^{o} \leftrightarrow (\exists x A(x))^{o}$

On the other hand, if we have, in the intuitionist calculus,

(2) $\vdash x \leftrightarrow y$, it is evident, from (2), that $\vdash x^{o} \equiv y^{o}$, where the meanings of the previous symbols are immediate.

Theorem 2.7 In $[B^{o}, \equiv, \rightarrow^{o}, \wedge, \vee, \neg^{o}, 0^{\circ}, 1_{\circ}]$, Peirce's law, $(x \rightarrow^{\circ} y) \rightarrow^{\circ} x \leq y$, is not valid.

Proof Just use the Boolean algebra of the subsets of the Euclidean plane and interpret "o" as an operator that makes each interior correspond to each set. (In this interpretation, $\equiv$ is the equality relation). All the postulates that define Boolean's concept of pre-algebra with constructibility operator are satisfied, but Peirce's law is not satisfied in the algebra of corresponding openings.

Problem 2.8 Given a Boolean pre-algebra, how many constructibility operators are there?

Definition 2.9 Given Boolean's pre-algebra with the constructibility operator ^{o}, $[B^{o}, \equiv, \rightarrow^{o}, \wedge, \vee, \neg^{o}, 0^{\circ}, 1_{\circ}]$, any Heyting algebra isomorphic to the algebra $[B^{o}, \equiv, \rightarrow^{o}, \wedge, \vee, \neg^{o}, 0^{\circ}, 1_{\circ}]/\equiv$ is called Heyting algebra associated with the Boolean pre-algebra considered.

Theorem 2.10 (Representation for Boolean algebras with the constructivity operator) All Boolean pre-algebra with the constructibility operator is associated with an open algebra.

Proof It is a consequence of all Heyting algebra being isomorphic to open algebra.

Definition 2.11 We call propositional Boolean pre-algebra to the Curry system [B, B*, $\rightarrow$, $\neg$] in which B $\neq \emptyset$, B* $\subseteq$ B, $\rightarrow$ is a binary operation in B and $\neg$ is an operator in B, with the following conditions being met:

(B_1) For all x, y, z $\in$ B, $(x \rightarrow (y \rightarrow z)) \rightarrow ((x \rightarrow y) \rightarrow (x \rightarrow z)) \in B^*$
(B_2) For all $x, y \in B$, $x \rightarrow (y \rightarrow x) \in B^*$
(B_3) For all $x, y \in B$, we have that $(\neg x \rightarrow \neg y) \rightarrow (y \rightarrow x)$.
(B_4) If $x \in B^*$ and $x \rightarrow y \in B^* \Rightarrow y \in B^*$

Definition 2.12 In the propositional Boolean pre-algebra [B, B*, $\rightarrow$, $\neg$]:

(1) $x \vee y$ abbreviates $\neg x \rightarrow y$.
(2) $x \wedge y$ abbreviates $\neg(x \rightarrow \neg y)$.
(3) $x \leftrightarrow y$ abbreviates $(x \rightarrow y) \wedge (y \rightarrow x)$.
(4) $\vdash x$ abbreviates $x \in B^*$
(5) $x \leq y$ abbreviates $\vdash x \rightarrow y$
(6) $x \equiv y$ abbreviates $x \leq y$ *and* $y \leq x$
(7) 0 abbreviates $x \wedge \neg x$

Definition 2.13 Let $[\mathrm{B},\ \mathrm{B}^*, \rightarrow, \neg]$ be a propositional Boolean pre-algebra in which the constructivity operator "o" is defined. So

(1) $\mathrm{B}^{\mathrm{o}} = \{x \mid \text{there exists } y \in \mathrm{B} \text{ such that } y^{\mathrm{o}} = x\}$
(2) $x \rightarrow^{\mathrm{o}} y$ abbreviates $(x \rightarrow y)^{\mathrm{o}}$
(3) $\neg^{\mathrm{o}}$ abbreviates $x \rightarrow^{\mathrm{o}} 0^{\mathrm{o}}$
(4) 1_{o} abbreviates $x^{\mathrm{o}} \rightarrow^{\mathrm{o}} x^{\mathrm{o}}$
(5) $\mathrm{B}^{*\mathrm{o}} = \{x \mid x \in \mathrm{B}^{\mathrm{o}} \text{and } x \equiv 1_{\mathrm{o}}\}$

Definition 2.14 With the notations of the previous definition, it is called propositional Boolean pre-algebra with constructor operator ^{o} to Boolean pre-propositional $[\mathrm{B},\ \mathrm{B}^*, \rightarrow, \neg]$, with the operator "o", the following postulates being valid:

(P_1) if $x \in \mathrm{B}$, then $x^{\mathrm{oo}} \equiv x^{\mathrm{o}}$
(P_2) if $x \in \mathrm{B}$, then $x^{\mathrm{o}} \leq x$.
(P_3) if $x, y \in \mathrm{B}$, then $(x \wedge y)^{\mathrm{o}} \equiv x^{\mathrm{o}} \wedge y^{\mathrm{o}}$
(P_4) if $x, y \in \mathrm{B}^{\mathrm{o}}$, then $x \wedge y, x \vee y \in \mathrm{B}^{\mathrm{o}}$
(P_5) if $x, y, z \in \mathrm{B}^{\mathrm{o}}$, then $(x \wedge z \leq y \Rightarrow z \leq x \rightarrow^{\mathrm{o}} y)$.

Theorem 2.15 With the notations of the definitions above, if $[\mathrm{B},\ \mathrm{B}^*, \rightarrow, \neg,\ ^{\mathrm{o}}]$ is a Boolean pre-algebra with the constructor operator ^{o}, $[\mathrm{B}, \rightarrow, \neg,\ ^{\mathrm{o}}]$, it is a pre-Heyting's algebra and $[\mathrm{B},\ \mathrm{B}^*, \rightarrow, \neg,\ ^{\mathrm{o}}]$ is Heyting's propositional pre-algebra.

Proof Immediate.

As the theory of C_n-algebras and propositional C_n-algebras was built, one could develop the theory of Boolean pre-algebras and Boolean propositional pre-algebras with the constructibility operator. The concepts of filter, deductive system, etc., would be defined, and their main properties would be studied. Of particular interest are Boolean's pre-algebras and propositional Boolean's pre-algebras with an operator, which are normal, i.e., those in which the condition is met: $1^0 \equiv 1_0 \equiv 1$.

3.3 Computability

Let T be any mathematical theory. Certain T propositions may be such that there are calculus methods that effectively demonstrate them. For example, in the theory of numerical equations or differential equations, the existence of solutions is often accompanied by processes by which these solutions can be "calculated" with the approximation you want. If this happens, the propositions in question are said to be computationally true. To express that proposition x is computationally true, we write x^*

k at T as Boolean's pre-algebra, the "*" operator enjoys precisely the
the constructibility operator. Boolean's notion of pre-algebra with the
ıs constitutes a generalization of the computability idea; see [8, 9].
ıe operator "o" could receive other similar interpretations.

3.4 Generalized Topological Spaces

Theorem 2.7 shows that Boolean's notion of pre-algebra with the constructibility operator generalizes the idea of topological space and extends the concept of Boolean algebra with the interior operator. In reality, it would be possible to define an open element, a border, a generalized continuous function, a compact space, etc. We would then arrive at a generalized topology, the purpose of which would be to study similar generalized topological spaces.

Certain peculiarities emerge in the "topological" theory of Boolean's pre-algebras with the operator o. For example, there are two types of open elements, according to the definitions below:

Definition 4.1 x is open if $x \equiv x^{o}$.

Definition 4.2 x is strictly open if $x = x^{o}$.

3.5 Duality

In Boolean's pre-algebra, we introduced the constructibility operator to obtain a Heyting pre-algebra; as it turned out, this structure generalizes the notion of topological space.

If we had proceeded dually, we would have dealt with Boolean pre-algebras with a "#" operator, and we would arrive at a Brower pre-algebra (dual pre-lattice of Heyting's pre-algebras), i.e., an absolute subtractive pre-lattice with the last element. From the topological point of view, the resulting structure would also be a generalization of the concept of topological space and, besides, of the broader idea of Boolean pre-algebra with a closure operator. Here, we only reproduce the definition of the new structure.

Definition 5.1 We call Boolean pre-algebras with the generalized closing operator, "#", all Boolean pre-algebra $[B, \equiv, \rightarrow, \wedge, \vee, ', 0, 1, {}^{\#}]$, with the operator "#", so so that postulates # 1 to # 5 follow, where $x \rightarrow^{\#} y$ abbreviates $(x \rightarrow y)^{\#}$, $\neg_{\#}x$ abbreviates $1^{\#} \rightarrow \# \ x$, $x^{\#}$ abbreviates the image of x by #, and $B^{\#}$ is the image of B by #:

($\#_1$) If $x \in B$, then $x^{\#\#} \equiv x^{\#}$.
($\#_2$) If $x \in B$, then $x^{\#} \leq x$.
($\#_3$) If $x, y \in B$, then $(x \wedge y)^{\#} \equiv x^{\#} \wedge y^{\#}$.
($\#_4$) If $x, y \in B^{\#}$, then $x \wedge y, x \vee y \in B^{\#}$.
($\#_5$) If $x, y, z \in B^{\#}$, then $(x \wedge z \leq y \Rightarrow z \leq x \rightarrow^{\#} y)$.

3.6 Heyting Pre-algebras with Ideality Operator

"Symmetric" situation of Boolean's pre-algebras with constructibility operator is obtained as follows:

Definition 6.1 Let [B, ≡, →, ∧, ∨, ¬, 0, 1] be a Heyting pre-algebra and suppose the operator I is defined in that pre-algebra. If $x \in$ B, the image of x by I will be denoted by x^{I}. So:

(1) x' abbreviates $(\neg x)^{\mathrm{I}}$
(2) $x \rightarrow {}^{\mathrm{I}}\, y$ abbreviates $(x \rightarrow y)^{\mathrm{I}}$
(3) B^{I} is the image of B by I

Definition 6.2 With the previous notations, the pre-algebra H with the operator I is called a Heyting pre-algebra with the ideality operator, I, or, still, a Heyting I-pre-algebra, if they occur. The postulates:

(I_1) $x^{\mathrm{II}} \equiv x^{\mathrm{I}}$
(I_2) $x \leq x^{\mathrm{I}}$
(I_3) $x \leq y \Rightarrow x^{\mathrm{I}} \leq y^{\mathrm{I}}$
(I_4) if $x, y \in \mathrm{B}^{\mathrm{I}}$, then $x \wedge y, x \vee y \in \mathrm{B}^{\mathrm{I}}$.
(I_5) if $x, y \in \mathrm{B}^{\mathrm{I}}$, then $x \rightarrow^{\mathrm{I}} y \equiv x' \vee y$
(I_6) if $x \in \mathrm{B}^{\mathrm{I}}$, then ($x \wedge x' \equiv 0^{\mathrm{I}}$and $x \vee x' \equiv 1^{\mathrm{I}}$)

Note. If $x \in$ B is a proposition, x^{I} can be interpreted as follows: x^{I} means that x is ideally or classically true.

Theorem 6.3 Let [B, ≡, →, ∧, ∨, ¬, 0, 1] be a Heyting pre-algebra with the ideality operator I. Then:

(1) $0^{\mathrm{I}}, 1^{\mathrm{I}} \in \mathrm{B}^{\mathrm{I}}$
(2) if $x, y \in \mathrm{B}^{\mathrm{I}}$, then $x \wedge^{\mathrm{I}} y,\ x \vee^{\mathrm{I}} y,\ x \rightarrow^{\mathrm{I}} y,\ x',\ y' \in \mathrm{B}^{\mathrm{I}}$

Proof Immediate.

Theorem 6.4 With the preceding notations, $[\mathrm{B}^{I}, \equiv, \rightarrow, \wedge, \vee, ', 0^{I}, 1^{I}]$ is a Boolean pre-algebra.

Proof Immediate.

Theorem 6.5 The postulate system of Heyting's I-pre-algebras is consistent.

Proof Let B be Heyting's algebra of all open sets in a topological space, and I the application that matches any open set of that algebra to the "smallest" simultaneously open and closed set of B that contains the open set in question. The algebra B, with the operator I, constitutes a Heyting I-pre-algebra.

Theorem 6.6 Let [B, ≡, →, ∧, ∨, ¬, 0, 1] be a Heyting pre-algebra. So:

(1) $0 \leq 0^{\mathrm{I}}$ and $1 \equiv 1^{\mathrm{I}}$.
(2) $x \vee y \leq (x \vee y)^{\mathrm{I}}$ and $x^{\mathrm{I}} \vee y^{\mathrm{I}} \leq (x \leq y)^{\mathrm{I}}$
(3) if $x \leq y$, then $x^{\mathrm{I}} \equiv y^{\mathrm{I}}$
(4) if $x, y \in \mathrm{B}^{\mathrm{I}}$, then $x \rightarrow^{\mathrm{I}} 0^{\mathrm{I}} \equiv 0^{\mathrm{I}}$
(5) if $x, y \in \mathrm{B}^{\mathrm{I}}$, then $x \leq y \Rightarrow x^{\mathrm{I}} \leq y$

(6) if $x, y \in B^I$, then $(x \vee y)^I \equiv x^I \vee y^I$
(7) if $x, y \in B^I$, then $(x \wedge y)^I \equiv x^I \wedge y^I$
(8) if $x \in B^I$, then $x^I \equiv x$
(9) $x^I \wedge y^I \equiv (x^I \wedge y^I)^I$ and $x^I \vee y^I \equiv (x^I \vee y^I)^I$

Proof Immediate.

Definition 6.7 Heyting B^I-pre-algebra is said to be strong if $x \in B^I$, carries $x \vee y \in B^I$, whatever $x, y \in B$.

Theorem 6.8 In a strong Heyting pre-algebra, if $x, y \in B$, then $x^I \vee y \in B^I$.

Proof Immediate.

Definition 6.9 A Heyting's I-pre-algebra $[B, \equiv, \rightarrow, \wedge, \vee, \neg, 0, 1, I]$ is said to be concrete if $[B, \equiv, \rightarrow, \wedge, \vee, \neg, 0, 1, I]$ is an open set algebra.

Theorem 6.10 Every Heyting pre-algebra $[B, \equiv, \rightarrow, \wedge, \vee, \neg, 0, 1, I]$ with the ideality operator is such that $B/\equiv$ is isomorphic to a concrete Heyting I-pre-algebra.

Problem 6.11 Starting with a pre-algebra of Boolean B, using the operator "^{o}", it is possible to define a pre-algebra of Heyting H. With the help of operator I, it is possible to define a pre-algebra of Boolean B*, starting from H. Under what conditions are B and B* isomorphic? Similarly, if we start with Heyting H's pre-algebra and the ideality operator I.

3.7 Intentional Operators

One more reason for the importance of non-monotonous operators in pre-lattices is the existence of intentional operators in logic; see [8].

In traditional logic, proposition composition has two basic categories: through extension operations and intentional operations. When composing a given proposition from others through an extensional operation, the truth value of the compound proposition depends exclusively on the truth values of the component propositions; the same is not true for intentional operations.

Typical examples of the extensive composition of propositions are those provided by the usual logical operations of conjunction, disjunction, material implication, and material equivalence.

The operator "I believe x", where x is a proposition we will represent by $x^{\#}$, is not extensional. If we have $x \leftrightarrow y$ (x is materially equivalent to y), it does not generally follow that $x^{\#} \leftrightarrow y^{\#}$. So, if we abbreviate the proposition "Misty is a song written by J. Burkee and E. Garner" by "x" and the proposition "J. Burkee and E. Garner wrote Johnny Mathis's most popular song", we have $x \leftrightarrow y$, but it does not necessarily follow that $x^{\#} \leftrightarrow y^{\#}$. Therefore, I can believe x without believing y, which would happen, for example, if I ignore that Johnny Mathis's most famous song is Misty.

To algebraically analyze the preceding operation, we start from Boolean's pre-algebra [B, ≡, →, ∧, ∨, ¬, 0, 1], in which the $^{\#}$ operator is defined. The system of postulates listed below can characterize the operator in question:

(C_1) if $x \in$ B, then $x^{\#\#} \equiv x^{\#}$.
(C_2) if $x, y \in$ B, then $(x \vee y)^{\#} \equiv x^{\#} \vee y^{\#}$.
(C_3) if $x, y \in$ B, then $(x \wedge y)^{\#} \equiv x^{\#} \wedge y^{\#}$.
(C_4) if $x \in$ B, then $x\prime^{\#} \leq x^{\#}$
(C_5) $1^{\#} \equiv 1$
(C_6) $0^{\#} \equiv 0$

Theorem 7.1 Let B be a Boolean pre-algebra with the # operator. We have:

(1) $(x \to y)^{\#} \equiv x^{\#} \to y^{\#}$
(2) $x^{\#} \wedge (x \to y)^{\#} \leq y^{\#}$
(3) $x^{\#} \vee 1^{\#\prime} \equiv 1^{\#}$
(4) $(x \wedge y)^{\prime\#} \leq x^{\#\,\prime} \wedge y^{\#\,\prime}$
(5) $(x \vee y)^{\prime\#} \leq x^{\#\,\prime} \wedge y^{\#\,\prime}$

Proof

(1) $x \to y$ is short for x' ∨ y. Thus,

$$(x \to y)^{\#} \equiv (x' \vee y)^{\#} \equiv x'^{\#} \vee y^{\#} \leq x^{\#\prime} \vee y^{\#} = x^{\#} \to y^{\#}$$

(2) $x^{\#} \wedge (x \to y)^{\#} \leq x^{\#} \wedge (x^{\#} \to y^{\#}) \leq y^{\#}$ because every Boolean pre-algebra is an implicative pre-lattice (defining $x \to y$ as an abbreviation for x' ∨ y).
(3) $(x \wedge y)^{\prime\#} \leq (x \wedge y)^{\#\prime} \equiv (x^{\#} \wedge y^{\#})' \equiv x^{\#\prime} \vee y^{\#\prime}$(3) and (5). They are immediate.

Theorem 7.2 With the same hypothesis as the previous theorem, $x \equiv y$ does not entail $x^{\#} \equiv y^{\#}$.

Proof Use Boolean's pre-algebra from theorem 2.6, identifying the operators "#" and "o".

Another system of postulates, analogous to the previous one and which could also be used to implicitly define one of the many meanings of the "I believe x" operator, is as follows:

(C'_1) if $x \in$ B, then $x^{\#\#} \equiv x^{\#}$.
(C'_2) if $x, y \in$ B, then $(x \wedge y)^{\#} \equiv x^{\#} \wedge y^{\#}$.
(C'_3) if $x, y \in$ B, then $(x \vee y)^{\#} \equiv x^{\#} \vee y^{\#}$.
(C'_4) if $x \in$ B, then $x^{\#\prime} \leq x^{\prime\#}$
(C'_5)$1^{\#} \equiv 1$

The last two structures are closely correlated with topological structures.

Problem 7.3 Obtain representation theorems for the last two types of structure.

Observation 7.4 Let [B, $\equiv$, $\rightarrow$, $\wedge$, $\vee$, $'$, 0, 1] be a Boolean pre-algebra, whose elements we will interpret as constituting propositions. If $x \in$ B, $[]x$, $\diamond x$ and Ix will represent, respectively, the propositions "x is necessary", "x is possible", and "x is impossible". Such operators are the primary modalities of traditional logic. It seems reasonable to assume that these operators are not monotonous concerning $\equiv$. In reality, two propositions can be materially equivalent, without both being necessary or possible. In this way, non-monotonous operators lead to a new way of looking at the logic of modalities. However, if propositions x and y are logically equivalent, then $\diamond x$ and $\diamond y$ will also be equivalent (the same with $[]x$ and []y, and Ix and Iy).

The basic problem, which motivates the study of non-monotonous operators in algebraic logic, is to be able to distinguish propositions x and y that are equivalent (in some sense, for example, materially equivalent), but that when the $^+$ (an intentional, modal, deontic, ...) operator, x^+ and y^+ are not always equivalent in the same sense as x and y. Moreover, the problem extends, in general, to non-monotonous operations of different arities.

References

1. Heyting, A.: Intuitionism: An Introduction (Second revised ed.). Amsterdam: North-Holland Publishing Co., (1966)
2. Heyting, Arend.: Die formalen Regeln der intuitionistischen Logik". Sitzungsberichte der preußischen Akademie der Wissenschaften, phys.-math. Klasse (in German): 42–56, 57–71, 158–169 (1930)
3. Troelstra, A.S., van Dalen, D.: Constructivism in mathematics: An introduction, North-Holland (1988)
4. Johnstone, P.: Stone spaces. Cambridge University Press (1982)
5. Dummett, M.: Elements of intuitionism, Clarendon Press (1977)
6. Bezhanishvili, G., Gehrke, M.: Lattices of intermediate and cylindric modal logics. J. Symb. Log. **76**(4), 1267–1282 (2011)
7. Goldblatt, R.: Mathematical modal logic: a view of its evolution. J. Appl. Log. **4**(4), 377–402 (2006)
8. Barros, C.M., Da Costa, N.C.A., Abe, J.M.: Tópico de Teoria dos sistemas ordenados, Vol II: sistemas de curry, Coleção documentos, Série Lógica e Teoria da Ciência, IEA-USP, 20, p. 132 (1995) ISSN 16799429
9. Abe, J.M., Nakamatsu, K., Akama, S.: Aspects of curry algebras, computability, constructibility, and topological spaces, frontiers in artificial intelligence and its applications, IOS Press, Amsterdam, **243**, 1342–1349, (2012). ISSN 0922-6389. https://doi.org/10.3233/978-1-61499-105-2-1342
10. Goldblatt, R.: Axiomatising the logic of computer programming. Lecture notes in computer science. Springer-Verlag, Berlin, Heidelberg, and New York, xi + 304 **130** (1982)
11. Jónsson, B., Tarski, A.: Boolean algebras with operators. Part I, Am. J. Math. **73**(4), 891–939 (1951). https://doi.org/10.2307/2372123

Chapter 4
P_n-Algebras

P_n-algebras represent a significant extension of algebraic structures designed to capture and formalize reasoning within non-classical logical systems, specifically those that allow paracomplete reasoning. These algebras are intimately connected to the study of deductive systems that depart from classical logical principles, particularly through their relationship to calculi like the P_1 calculus, developed by da Costa and Marconi, which provides a foundational framework for paracomplete logic; see da Costa & Marconi [1].

The study of P_n-algebras focuses on generalizing classical algebraic structures, such as Boolean and Heyting algebras, to accommodate the unique logical operations in paracomplete systems. In classical logic, where every proposition must be true or false, deductive systems correspond neatly to Boolean algebra structures. However, in paracomplete logics, such as those studied by da Costa, propositions may remain indeterminate or fail to behave according to the Principle of Excluded Middle. P_n-algebras are designed to model these situations algebraically, allowing us to formalize reasoning in systems where paracompleteness does not lead to trivial conclusions.

This chapter explores the core properties of P_n-algebras, focusing on their relationship with deductive systems and homomorphisms. Deductive systems formalize inference rules in logical calculi, and P_n-algebras provide an algebraic interpretation of these rules, translating syntactic operations on formulas into algebraic operations on elements. By studying homomorphisms between P_n-algebras, we can examine how logical equivalences and transformations between different reasoning systems are preserved and interpreted within algebraic frameworks.

In the case of P_1, da Costa and Marconi [1] showed that a robust algebraic framework can be used to address these meta-logical properties, such as the study of soundness and completeness of the logical calculi P_n, thus linking the syntactic properties of the calculus to their algebraic counterparts in P_n-algebras.

This investigation into P_n-algebras provides a deeper understanding of the structural properties of paracomplete logics and opens avenues for further exploration of homomorphisms and algebraic interpretations in more complex paracomplete

J. M. Abe, *An Introduction to Curry Systems*, Intelligent Systems Reference Library 280, https://doi.org/10.1007/978-3-031-97810-4_4

systems. By extending classical results on deductive systems into the non-classical domain, we aim to shed light on the interplay between algebraic structures and logical calculi in the context of paracomplete reasoning.

We move on to the study of P_n-algebras. Some of its main properties and relationship with deductive systems and homomorphisms are studied. We present an application of this investigation in treating issues related to the soundness and completeness of the P_1 calculus by da Costa and Marconi [1].

4.1 Calculus P_n, $1 \leq n < \omega$

In one of its possible formulations, the Principle of Excluded Middle says that, of two contradictory propositions, A and $\neg A$, one is true. Paracomplete logic can serve as a logic underlying theories that contain a false proposition, and its negation is also false. Therefore, paracomplete logic derogates the Principle of Excluded Middle. In this sense, paracomplete logic comprises several types of logic, such as intuitionist logic and many-valued logic.

A motivation for paracomplete logic is that the classic condition "of two contradictory propositions, one is true" does not always reflect our intuitions. For example, if P is a vague predicate and x is an individual from the vagueness frontier, we can assume that sometimes Px and $\neg$Px are false.

In [1], da Costa and Marconi built a hierarchy P_n ($1 \leq n < \omega$) of paracomplete propositional calculus. These calculus are 'dual' of the paraconsistent propositional logics, C_n, $1 \leq n < \omega$ (cf. §1, Chap. 2).

We also observed that the paracomplete P_n logics are related to the ideas that underlie fuzzy logic.

Initially, we will describe the propositional calculus for the complete P_1. The primitive symbols of P_1 are $\rightarrow$ (implication), $\bigvee$ (disjunction), $\wedge$ (conjunction), $\neg$ (negation), parentheses, and propositional variables. The equivalence symbol $\leftrightarrow$ is defined in the usual way. The concepts of formula and sub-formula are defined commonly. A, B, and C denote formulas, and A * is short for $A \vee \neg A$.

They are postulates (axiom schemes and primitive rules) of P_1.

The postulates of P_1 (axiom schemes and Inference rule) are as follows: A, B, and C are formulas, and A^* is short for $A \vee \neg A$.

$(\to_1) A \to (B \to A)$

$(\to_2)\ (A \to B) \to ((A \to (B \to C)) \to (A \to C))$

$(\to_3) \dfrac{A, A \to B}{B}$ (MP (Modus Ponens))

$(\wedge_1) A \wedge B \to A$

$(\wedge_2)\ A \wedge B \to B$

$(\wedge_3) A \to (B \to (A \to B))$

$(\vee_1) A \to (A \vee B)$

$(\vee_2) B \to (A \vee B)$

$(\vee_3)\ (A \to C) \to ((B \to C) \to (A \vee B \to C))$

$(\neg_1) A^* \to ((A \to B) \to ((A \to \neg B) \to \neg A))$

$(\neg_2) A^* \wedge B^* \to (A \to B)^* \wedge (A \wedge B)^* \wedge (A \vee B)^* \wedge (\neg A)^*$

$(\neg_3) \neg(A \wedge \neg A)$

$(\neg_4) A \to (\neg A \to B)$

$(\neg_5)\ A \to \neg\neg A$

The intuitive justification for the postulates above is analogous to the justification for the postulates of calculus C_1, seen in §1, Chap. 2, with adaptations.

Definition 1.1 Let A be a formula. We define A^n inductively as follows:

(a) $A^1 =_{Def} A^*$

(b) $A^n = A^* \wedge A^{**} \wedge \ldots \wedge A^{\overbrace{**\ldots*}^{n-times}}$

Definition 1.2 Let A be a formula. We define $A^{(n)}$ inductively as:

(a) $A^{(1)}$ abbreviates A^1
(b) $A^{(n)}$ abbreviates $A^{(n)} \wedge A^{(n+1)}$.

The calculus P_n, $1 \leq n < \omega$, has the same language as P_1 and the same postulates, except the axiom schemes $(\neg_1)$ and $(\neg_2)$, which are replaced, respectively, by:

$$(\neg_{n1}) A^{(n)} \to ((A \to B) \to ((A \to \neg B) \to \neg A))$$
$$(\neg_{n2}) A^{(n)} \wedge B^{(n)} \to (A \to B)^{(n)} \wedge (A \wedge B)^{(n)} \wedge (A \vee B)^{(n)} \wedge (\neg A)^{(n)}$$

The calculus P_ω is obtained from P_1 by suppressing the axiom schemes $(\neg_1)$ and $(\neg_2)$.

Theorem 1.3 *In P_1, all the theorems and rules of the classical positive calculus are valid.*

Proof It is immediate, considering the postulates of P_1.

Theorem 1.4 *In P_1, the following schemes are valid*:

(1) $\vdash A \wedge \neg A \rightarrow B$
(2) $\vdash A \vee (A \rightarrow B)$
(3) $\vdash A^* \wedge (\neg\neg A \rightarrow A)$
(4) $\vdash A \rightarrow ((A \rightarrow B) \wedge (A \rightarrow \neg B) \rightarrow \neg A))$
(5) $\vdash A \wedge \neg A \leftrightarrow \neg B \wedge \neg B$.

Proof Immediate.

Theorem 1.5 *In P_1, the schemes are not valid*:

(1) $A \vee \neg A$
(2) $\neg(A \vee B) \leftrightarrow \neg A \wedge \neg B$
(3) $\neg(A \wedge B) \leftrightarrow \neg A \vee \neg B$
(4) $\neg\neg A \rightarrow A$
(5) $\neg\neg A \leftrightarrow A$
(6) $(A \rightarrow B) \rightarrow (\neg B \rightarrow \neg A)$
(7) A^{**}.

Proof Immediate.

Definition 1.6 In P_1, we define a "strong negation", $\neg^*$, as follows:

$$\neg^* A =_{\text{Def}} A \rightarrow (B \wedge \neg B)$$

where B is a fixed formula.

Theorem 1.7 *In P_1, strong negation has all the properties of classical negation.*

Proof Immediate.

Corollary 1.7.1 *In P_1, the schemes are valid*:

(1) $(A \rightarrow B) \wedge ((A \rightarrow \neg^* B) \rightarrow \neg^* A)$
(2) $A \rightarrow (\neg^* A \rightarrow B)$
(3) $A \vee \neg^* A$.

Proof Due to the previous theorem.

Theorem 1.8 *If P_0 denotes the classical propositional calculus, then each calculus of the sequence P_0, P_1, P_2, ..., P_n, ..., P_ω is strictly stronger than each one of the preceding.*

Proof Analogous to the case of C_n -Calculus.

4.2 The P_1-Algebra

In this paragraph, we study the algebraic version of the propositional calculus for the complete P_n, $1 \leq n < \omega$. As a consequence, we obtain P_n-algebras, which algebraize the P_n system. The P_n system negation is incompatible with the basic equivalence relation between formulas; therefore, P_n-algebras constitute Curry algebras.

Definition 2.1 A P_1-algebra is a classic implicative pre-lattice $[R, \equiv, \leq, \wedge, \vee, \rightarrow, ']$, with a greater element and with an operator'satisfying:

$(P_1) x \leq x''$
$(P_2)\ (x \wedge x')' \equiv 1$
$(P_3) x^* \wedge y^* \leq (x \rightarrow y)^*$
$(P_4) x^* \wedge y^* \leq (x \wedge y)^*$
$(P_5) x^* \wedge y^* \leq (x \vee y)^*$
$(P_6) x^* \leq (x')^*$
$(P_7) x^* \leq (x \rightarrow y) \rightarrow ((x \rightarrow y') \rightarrow x')$
$(P_8) x \leq (x' \rightarrow y)$

Consider the propositional paracomplete calculus P_1. Let F be the set of formulas for P_1. Let us define operations $\rightarrow$, $\wedge$, $\vee$, and $'$, in F, respectively, by:

1. $x \rightarrow y =_{\text{Def}} x \rightarrow y$
2. $x \wedge y =_{\text{Def}} x \wedge y$
3. $x \vee y =_{\text{Def}} x \vee y$
4. $x' =_{\text{Def}} \neg x$

Let us also define the relation $\equiv$ over F as:

$$x \equiv y =_{\text{Def}} \vdash x \leftrightarrow y$$

Theorem 2.2 *With the notations and definition above, the Curry system* $[F, \equiv, \rightarrow, \wedge, \vee]$ *is a classic implicative pre-lattice with a first element.*

Proof A consequence of Theorems 2.2 and 2.4, Chap. 2.

Theorem 2.3 *The Curry system* $[F, \equiv, \rightarrow, \wedge, \vee, ']$ *is a* P_1*-algebra.*

Proof In effect, it is a consequence of the axioms $(\neg 1)$–$(\neg 5)$ of the P_1 calculus.

Theorem 2.4 *Let* $[S, \equiv, \rightarrow, \wedge, \vee, ']$ *be a* P_1*-algebra. The operator' is generally not monotonous concerning the fundamental relation.*

Proof In the model $[F, \equiv, \rightarrow, \wedge, \vee, ']$ considered in the previous theorem, the operator' is not monotonous concerning the fundamental relation $\equiv$ between P_1 formulas.

Corollary 2.5 *Every P_1-algebra is a Curry algebra.*

Proof A consequence of the previous theorem.

Theorem 2.6 *A P_1-algebra has a maximum element and is distributive.*

Proof Immediate.

Definition 2.7 Let [B, $\equiv$, $\rightarrow$, $\wedge$, $\vee$, $'$] be a P_1-algebra. For any element x of S, $\neg^* x$ abbreviates $x \rightarrow (y \wedge y')$, where y is a fixed element of S.

Theorem 2.8 *In a P_1-algebra, the operator $\neg^*$ defined above has all the properties of Boolean complementation,* i.e., $x \vee \neg^* x \equiv 1$ and $x \wedge \neg^* x \equiv 0$.

Proof Similar to theorem 2.13, Chap. 2.

Theorem 2.9 *In a P_1-algebra, the structure composed of the underlying set and the operations $\wedge$, $\vee$ and $\neg^*$ is a Boolean pre-algebra.*

Proof Consequence of the previous theorem.

Observation 2.10 By Theorem 6.3 (Chap. 1), we can pass the Boolean pre-algebra obtained in the previous theorem to the quotient, obtaining a Boolean algebra in the usual sense. We thus obtain relevant representation theorems.

Definition 2.11 Let [B, $\equiv$, $\rightarrow$, $\wedge$, $\vee$, $'$] be a P_1-algebra and B/$\equiv$ the Boolean algebra obtained by passing to the quotient the Boolean pre-algebra [B, $\equiv$, $\rightarrow$, $\wedge$, $\vee$, $'$] by the fundamental relation $\equiv$. Boolean algebra is isomorphic to the quotient B/$\equiv$ algebra is called Boolean algebra associated with the P_1-algebra in question.

Theorem 2.12 *Every P_1-algebra is associated with a reduced field of sets.*

Proof Similar to Theorem 2.17, Chap. 1.

Theorem 2.13 *All P_1-algebra is associated with the field of the sets, both open and closed, in a compact and totally disconnected Hausdorff space* [2–4].

Proof Similar to the previous theorem.

Problem 2.14 How many non-isomorphic P_1-algebras are associated with the same Boolean algebra?

4.3 The P_n-Algebras, $1 \le n < \omega$

Most of the results of P_1-algebras extend without difficulty to P_n-algebras, $1 \le n < \omega$. Here, we present only P_n-algebras' definitions and will mention some results.

Definition 3.1 Let [B, ≡, →, ∧, ∨, ′] be a P_n-algebra. If $x \in B$, x^1 abbreviates x.

x^n abbreviates $x' \wedge x'' \wedge, \ldots, x''^{\cdots, \prime}$, where, in the latter expression, the symbol ’ appears n times.

Definition 3.2 In the preceding definition conditions,

$x^{(1)}$ abbreviates x^1

$x^{(n)}$ abbreviates x^n

Definition 3.3 A P_n-algebra is an absolute implicative pre-lattice [B, ≡, → , ∧, ∨, ′], with a larger element, in which an operator’ is defined as satisfying the following conditions:

$(P_1) x \le x''$
$(P_2)\ (x \wedge x')' \equiv 1$
$(P_3) x^{(n)} \wedge y^{(n)} \le (x \to y)^{(n)}$
$(P_4) x^{(n)} \wedge y^{(n)} \le (x \wedge y)^{(n)}$
$(P_5) x^{(n)} \wedge y^{(n)} \le (x \vee y)^{(n)}$
$(P_6) x^{(n)} \le (x')^{(n)}$
$(P_7) x^{(n)} \le (x \to y) \to ((x \to y') \to x')$
$(P_8) x \le (x' \to y)$

Theorem 3.4 *All P_n-algebra, $1 \le n < \omega$, is a classical and distributive implicative pre-lattice.*

Proof Immediate.

4.4 The P_ω-Algebra

Definition 4.1 A P_ω-algebra is an absolute implicative pre-lattice, with the last element, in which an operator’ is defined, satisfying the following conditions:

$(P_1) x \le x''$
$(P_2)\ (x \wedge x')' \equiv 1$

Theorem 4.2 *P_ω-algebras are classical implicative pre-lattices and have, in general, no last element.*

Proof They are left as an exercise.

Theorem 4.3 *If $m \leq n$, any P_m-algebra is also a P_n-algebra ($m, n = 0, 1, 2, \ldots, \omega$).*

Proof Immediate.

Theorem 4.4 *Every Boolean pre-algebra is a P_n-algebra, $1 \leq n < \omega$.*

Proof Immediate.

4.5 The Propositional P_n-Algebras, 1 ≤ n < ω

We now introduce the concept of propositional P_n-algebra, analogous to that of propositional C_n-algebra (studied in Chaps. 2 and 5). This concept also generalizes P. Rosenbloom's propositional Boolean algebra. We observed, in passing, that we could deal with the general theory of propositional P_n-algebras, $1 \leq n < \omega$, but here we limit ourselves only to propositional P_1-algebras.

Definition 5.1 A propositional P_1-algebra is a Curry system $[R, R^*, \equiv, \rightarrow, \wedge, \vee, \neg]$, where $R^* \subseteq R \subseteq R$ (and $R \neq \emptyset$), $\rightarrow, \wedge, \vee$ are binary operations in R and $\neg$ is an operator in R, the following postulates are satisfied $x^* \rightarrow (y \vee y')$.

$(B_1) x, y \in R \Rightarrow x \rightarrow (y \rightarrow x) \in R^*$

$(B_2) x, y \in R \Rightarrow (x \rightarrow y) \rightarrow ((x \rightarrow (y \rightarrow z)) \rightarrow (x \rightarrow z)) \in R^*$

$(B_3) x, y \in R \Rightarrow x \in R^*$ and $x \rightarrow y \in R^* \Rightarrow y \in R^*$

$(B_4) x, y \in R \Rightarrow x \wedge y \rightarrow x \in R^*$

$(B_5) x, y \in R \Rightarrow x \wedge y \rightarrow y \in R^*$

$(B_6) x, y \in R \Rightarrow x \rightarrow (y \rightarrow (x \rightarrow y)) \in R^*$

$(B_7) x, y \in R \Rightarrow x \rightarrow (x \vee y) \in R^*$

$(B_8) x, y \in R \Rightarrow y \rightarrow (x \vee y) \in R^*$

$(B_9) x, y \in R \Rightarrow (x \rightarrow z) \rightarrow ((y \rightarrow z) \rightarrow (x \vee y \rightarrow z)) \in R^*$

$(B_{10}) x, y \in R \Rightarrow x \rightarrow \neg\neg x \in R^*$

$(B_{11}) x, y \in R \Rightarrow \neg(x \wedge \neg x) \in R^*$

$(B_{12}) x, y \in R \Rightarrow y \rightarrow ((x \rightarrow y) \rightarrow ((x \rightarrow \neg y) \rightarrow \neg x)) \in R^*$

$(B_{13}) x, y \in R \Rightarrow x^* \wedge y^* \rightarrow (x \rightarrow y)^* \in R^*$

$(B_{14}) x, y \in R \Rightarrow x^* \wedge y^* \rightarrow (x \wedge y)^* \in R^*$

$(B_{15}) x, y \in R \Rightarrow x^* \wedge y^* \rightarrow (x \vee y)^* \in R^*$

$(B_{16}) x, y \in R \Rightarrow x^* \rightarrow (\neg x)^* \in R^*$

Let [R, R*, $\equiv$, $\rightarrow$, $\wedge$, $\vee$, $\neg$] be a propositional P_1-algebra. Suppose that $x, y \in$ R. Then $x \leftrightarrow y$ abbreviates $(x \rightarrow y) \wedge (y \rightarrow x)$. If $x \in$ R*, we also write $\vdash x.x \equiv y$ is also an abbreviation for $\vdash x \rightarrow y$ and $x \leq y$ for $\vdash x \rightarrow y$. It can be shown that $\equiv$ is an equivalence relation and that $\leq$ is a quasi-order. It is easily shown that there are elements 0 and 1 such that $0 \leq x$ and $x \leq 1$, for all x.

It follows that if [R, R*, $\equiv$, $\rightarrow$, $\wedge$, $\vee$, $'$] is a propositional P_1-algebra, if we define $\equiv$ as above. If 0 is a first element of this algebra, and' abbreviates $\neg$, then [R, $\equiv$, $\rightarrow$, $\wedge$, $\vee$, $'$] is a P_1-algebra.

Conversely, if [R, $\equiv$, $\rightarrow$, $\wedge$, $\vee$, $'$] is a P_1-algebra, defining R* as the set of elements x of R* such that $x \equiv \mathrm{y} \rightarrow \mathrm{y}$, for a fixed y of R and representing the operator ' by $\neg$, [R, R*, $\equiv$, $\rightarrow$, $\wedge$, $\vee$, $\neg$] is a propositional P_1-algebra.

The proof of the equivalence of such structures is similar to what was done in §5, Chap. 2, for the case of propositional P_1-algebras.

Relevant results to the purposes of this book are that the propositional P_1-algebra, [R, R*, $\equiv$, $\rightarrow$, $\wedge$, $\vee$, $\neg$], the operator $\neg$ is not monotonous relative to $\equiv$ (see definition above).

Also, if in propositional P_1-algebra [R, R*, $\equiv$, $\rightarrow$, $\wedge$, $\vee$, $\neg$] we define the $\mathrm{R}' = \{x \in \mathrm{R} \mid \vdash x^*\}$ and $\mathrm{R}^{*\prime} = \{x \in \mathrm{R} \mid \vdash x^* \wedge x\}$, so the system [R$'$, R$^{*\prime}$, $\equiv$, $\rightarrow$, $\wedge$, $\vee$, $\neg$] constitutes a propositional Boolean algebra. If we consider P_1-algebra [R, $\equiv$, $\rightarrow$, $\wedge$, $\vee$, 1,$'$] and $\mathrm{R}^* = \{x \in \mathrm{R} \mid \vdash x^* \equiv 0\}$, then [R*, $\equiv$, $\wedge$, $\vee$, $'$] is a Boolean algebra. These results show that the elements "well behaved" (i.e., those that satisfy the Principle of Excluded Middle) in P_1-algebras or propositional P_1-algebras) behave in a "classic" way.

4.6 Deductive Systems

Definition 6.1 Let [R, $\equiv$, $\rightarrow$, $\wedge$, $\vee$, 1, $'$, 0, 1] be a P$_1$-algebra and M a subset of P. A finite sequence $x_1, x_2, \ldots, x_\mathrm{n}$ of elements of R says if a deduction of x ($x\in$ R) from hypotheses M if:

(i) For all i, i, $1 \leq \mathrm{i} \leq \mathrm{n}$, $x_\mathrm{i} \in$ M, $x_\mathrm{i} = 1$ or there are j, k < i such that x_k is $x_\mathrm{j} \rightarrow x_\mathrm{i}$.
(ii) x_n is x.

If there is a deduction of x from the hypotheses M, x is said to be a consequence of M, and we write M $\vdash x$. The set of consequences of M is symbolized by M$^{\#}$. If M = M$^{\#}$, it is said to be a deductive system or theory.

4.8 Soundness and Completeness of P_1

Theorem 8.1 *If the co-nucleus of a homomorphism f of F (the class of the formulas of P_1) in a P_1-algebra is an ultrafilter, then: (A, B $\in$ F).*

(1) f (A) results in f(¬A) ≡ 0.
(2) f (A) not equivalent to f (¬A) results in f (¬A) not equivalent to f(¬¬A).
(3) If f (A) is not equivalent to f (¬A) and f(B) is not equivalent to f (¬B), then f (A → B) not equivalent to f(¬(A → B)) and f(A ⋀ B) not equivalent to f(¬(A ⋀ B)) and f(A ⋁ B) not equivalent to f(¬(A ⋁ B)).
(4) f(¬(A ⋀ ¬A)) ≡ 1.

Proof Immediate.

Definition 8.2 Let F be the class of formulas of P_1 and f be a homomorphism of F in a P_1-algebra. We denote by f ⊨ Γ, where Γ is a subset of F, if for each A ∈ F, f (A) ≡ 1. Γ ⊨ A means that for any homomorphism f of F in an arbitrary P_1-algebra, if f ⊨ Γ, then f (A) ≡ 1.

Theorem 8.3 *(Soundness) If A is a provable formula of P_1, then f (A) ≡ 1 for any homomorphism f of F in an arbitrary P_1-algebra (i.e., ⊨ A).*

Proof By length induction.

Theorem 8.4 *Let U be an ultrafilter in F. So there is a homomorphism f of P_1 in 2 (see Lemma 6.2, Chap. 2) such that the co-nucleus of f is U.*

Proof Immediate.

Theorem 8.5 *(Completeness) Let A ∈ F. Suppose that f (A) ≡ 1 for any homomorphism of P_1 in an arbitrary P_1-algebra. So, A is a provable formula of P_1.*

Proof Indeed, if A is not provable, then A is not equivalent to 1, and therefore ¬A is also not equivalent to 0. Thus, there is an ultrafilter U in F that contains A. By the previous theorem, there is a homomorphism f of P_1 to 2 and, therefore, f (A') ≡ 1. It follows that f (A) ≡ 0, which is a contradiction; see Abe [2].

References

1. da Costa, Newton C. A., and Marconi, Diego. "A note on paracomplete logic." Atti della Accademia Nazionale dei Lincei. Classe di Scienze Fisiche, Matematiche e Naturali. Rendiconti Lincei. Matematica e Applicazioni 80.7-12 (1986): 504-509.
2. Abe, J.M.: A note on curry algebras, bulletin of the section of logic. Polish Academy of Science **16**(4), 151–158 (1987)

Chapter 5
N_n-Algebras

The non-alethic logics N_n emerge from clarifying the duality between C_n paraconsistent systems, as formulated by Newton da Costa, and their complementary paracomplete counterparts. While C_n systems were designed to deal with contradictions without incurring triviality, thus breaking with the classical Principle of Non-contradiction [1], P_n systems reject the Principle of Excluded Middle, according to which, for any proposition p, it must be true that either p or $\neg p$; see [2]. Paracomplete logics, therefore, admit the possibility that both p and $\neg p$ are simultaneously false in contexts of uncertainty or imprecision.

The discovery of this duality paved the way for constructing logical systems that simultaneously integrate paraconsistent and paracomplete properties. These hybrid systems, called non-alethic logics, break with the traditional conception of truth (aletheia, from the Greek), rejecting its centrality as a fundamental criterion for logical evaluation. Instead of being based on the truth/falsehood dichotomy, non-alethic logics propose a more flexible approach capable of formalizing reasoning in contexts where contradictions and indeterminacies coexist; see da Costa [2].

In this context, introducing N_n propositional systems significantly expanded the applicability of non-alethic logics. These systems formalize the integration between paraconsistent and paracomplete characteristics, allowing the simultaneous treatment of contradictory, indeterminate, or both propositions.

It is important to highlight that, although they present similarities in their non-classical treatment of negation and contradiction, paraconsistent and paracomplete logics are conceptually and mathematically independent. The former focuses on managing contradictions without the system collapsing into triviality, while the latter focuses on rejecting the ontological completeness implied by the Principle of Excluded Middle. Thus, paraconsistent logics are not paracomplete, and vice versa.

The synthesis of these approaches in non-alethic systems represents a relevant advance for several reasons. First, it expands the conceptual space for constructing logical systems outside the classical structures of functional truth. Second, such

J. M. Abe, *An Introduction to Curry Systems*, Intelligent Systems Reference Library 280, https://doi.org/10.1007/978-3-031-97810-4_5

logics are particularly suitable for specific AI, Intelligent Systems, Automation, and Robotics applications.

In addition, several non-alethic systems have been developed, such as annotated logics, which allow more sophisticated evaluations of propositions by assigning annotations or additional evidence to their logical values. This mechanism enables the formal modeling of uncertainties, imprecision, and contradictions in a way that classical logic cannot accommodate; see Abe et al. [3]. Integrating paraconsistent and paracomplete components gives non-alethic logics a versatile and robust framework for dealing with the complexities inherent in reasoning in formal systems applied to the real world.

In summary, non-alethic logics represent a substantial development in non-classical logic by synthesizing paraconsistent and paracomplete reasoning mechanisms into a unified framework. With the introduction of N_n systems, these logics offer new approaches for managing contradictions and paracompleteness, contributing to the expansion of the boundaries of formal logic and providing effective tools for dealing with complex, inconsistent, or paracomplete environments.

5.1 The Calculus N_n ($1 \leq n \leq \omega$)

In [2] N_n ($1 \leq n < \omega$) non-alethic propositional systems were introduced. In this section, we summarize a family of N_n non-alethic propositional N_n ($1 \leq n < \omega$). For each n, $1 \leq n < \omega$, we have a different calculus represented by N_n. N_n ($1 \leq n < \omega$) non-alethic calculus are 'hybrid' calculus of C_n ($1 \leq n < \omega$) and P_n ($1 \leq n < \omega$) calculus. Thus, such calculus has the following characteristics:

(1) The Priciple of Excluded Middle, in the form $A \vee \neg A$ is not generally valid;
(2) The Principle of Non-contradiction, in the form $\neg(A \wedge \neg A)$ is not generally valid;
(3) We cannot deduce any formula B from two contradictory propositions, A and $\neg A$.
(4) There are formulas A and $\neg A$ that are both true.
(5) There are formulas A and $\neg A$ that are both false.
(6) The calculus must contain the most important schemes and inference rules of the classical propositional calculus compatible with conditions (1)–(5) above.

The language of calculus N_n ($1 \leq n < \omega$) is the same for all of them; Let us denote by L. The language L is a language of the first order whose primitive symbols are the following:

(1) Propositional variables: an enumerable infinite set of propositional variables.
(2) Logical connectives: $\neg$ (negation), $\wedge$ (conjunction), $\vee$ (disjunction), and $\rightarrow$ (implication).
(3) Auxiliary symbols: parentheses and commas.

Formulas and other usual syntactic concepts are usually defined. In L, we put:

Definition 1.1 Let A be a formula. We have to A^{o} abbreviates $\neg(A \wedge \neg A)$. A^{i} abbreviates $A^{o \cdots o}$, where the symbol ^{o} appears i times, $i \geq 1$. (Thus, A^1 is A^{o}). We also write $A^{(i)}$ for $A^1 \wedge A^2 \wedge A^3 \wedge \ldots \wedge A^i$.

Definition 1.2 Let A be a formula. We have that $A^{\#}$ abbreviates $A \vee \neg A$. We also write $A^{[i]}$ for $A^{\#} \wedge A^{\#\#} \wedge A^{\#\#\#} \wedge \ldots \wedge A^{\#n}$, where #n means that the symbol # appears n times.

Definition 1.3 We write.

(1) $(A \leftrightarrow B)$ for $(A \rightarrow B) \wedge (B \rightarrow A)$ (equivalence)
(2) $\neg^{*}A$ for $A^{(n)} \wedge \neg A$ (strong negation).

The postulates (axiom schemes and inference rule) of N_n $(1 \leq n < \omega)$ are as follows: A, B, and C denote any formulas.

$(\rightarrow_1)$ $A \rightarrow (B \rightarrow A)$
$(\rightarrow_2)$ $(A \rightarrow (B \rightarrow C)) \rightarrow ((A \rightarrow B) \rightarrow (A \rightarrow C))$
$(\rightarrow_3)$ $\frac{A, A \rightarrow B}{B}$ (MP (Modus Ponens))
$(\wedge_1)$ $A \wedge B \rightarrow A$
$(\wedge_2)$ $A \wedge B \rightarrow B$
$(\wedge_3)$ $A \rightarrow (B \rightarrow (A \wedge B))$
$(\vee_1)$ $A \rightarrow A \vee B$
$(\vee_2)$ $B \rightarrow A \vee B$
$(\vee_3)$ $(A \rightarrow C) \rightarrow ((B \rightarrow C) \rightarrow ((A \vee B) \rightarrow C))$
$(\neg_{10})$ $A^{[n]} \wedge B^{(n)} \rightarrow ((A \rightarrow B) \rightarrow ((A \rightarrow \neg B) \rightarrow \neg A)$
$(\neg_4)$ $(A^{(n)} \wedge B^{(n)}) \rightarrow ((A \wedge B)^{(n)} \wedge (A \vee B)^{(n)} \wedge (A \rightarrow B)^{(n)})$
$(\neg_9)$ $(A^{[n]} \wedge B^{[n]}) \rightarrow ((A \wedge B)^{[n]} \wedge (A \wedge B)^{[n]} \wedge (A \rightarrow B)^{[n]})$
$(\neg_{11})$ $A^{(n)} \rightarrow (A \rightarrow \neg\neg A) \wedge (A \rightarrow (\neg A \rightarrow B))$
(N) $A^{(n)} \vee A^{[n]}$

The N_ω calculus is obtained from N_n by suppressing the axiom schemes $(\neg_{10})$, $(\neg_4)$, $(\neg_9)$, $(\neg_{11})$, $(\neg_{12})$, (N).

Theorem 1.4 In N_n $(1 \leq n < \omega)$, all valid schemes and rules of classical positive propositional logic are valid. In particular, the deduction theorem is valid N_n $(1 \leq n < \omega)$. N_n $(1 \leq n < \omega)$ contains the positive intuitionist logic.

Proof An immediate consequence of the postulates. Since Peirce's Law is valid in N_n $(1 \leq n < \omega)$, they contain the positive intuitionistic logic.

Theorem 1.5 *In N_n $(1 \leq n < \omega)$ we have:*
$\vdash \neg A^{(n)} \rightarrow (\neg A)^{(n)}$

Since the N_n $(1 \leq n < \omega)$ calculus was built from the C_n $(1 \leq n < \omega)$ calculus and the P_n $(1 \leq n < \omega)$ calculus, the intuitive meanings of the expressions $A^{(n)}$ and $A^{[n]}$ are the same as those mentioned. It should be noted, then, that the calculus N_n $(1 \leq n < \omega)$ has all the characteristics of C_n and P_n, being paraconsistent and paracomplete logics simultaneously. Furthermore, in these calculi, the connectives $\rightarrow$

, $\wedge$, $\vee$ and $\neg^*$ have all the properties of classical implication, conjunction, disjunction, and negation, respectively, which makes the classical propositional calculus contained in N_n.

Theorem 1.6 *In* N_n $(1 \leq n < \omega)$, *the following scheme is not valid*:
$A^o \wedge \neg A$

Theorem 1.7 *In* N_n $(1 \leq n < \omega)$, *the following schemes are not valid*:

1. $A \vee \neg A$
2. $\neg(A \vee B) \leftrightarrow (\neg A \wedge \neg B)$
3. $\neg(A \wedge B) \leftrightarrow (\neg A \vee \neg B)$
4. $\neg\neg A \rightarrow A$
5. $(A \wedge \neg A) \leftrightarrow (B \wedge \neg B)$.

If N_0 denotes the classical propositional calculus, then the hierarchy N_n $(0 \leq n \leq \omega)$ or N_0 , N_1, N_2, ..., N_n, ..., N_ω is such that N_{i+1} is strictly weaker than N_i for all ($1 \leq i \leq \omega$). N_ω is the weakest calculus in the hierarchy. It is worth noting that we can extend N_i to higher-order logics and strong theories that can be formalized in these logics.

Also, these logics have double-valued semantics, which are sound and complete. Let us consider only the N_1 calculus in order to fix ideas. N_1 is a paraconsistent and paracomplete calculus, that is, non-alethic; thus, we can use it to manipulate a set of inconsistent and paracomplete formulas without immediate trivialization (by this, we mean that not all formulas are theorems).

As we noted, when we are only working with formulas that satisfy the Principle of Non-contradiction, N_1 reduces to C_1, and if we are considering formulas that only satisfy the Principle of Excluded Middle, N_1 reduces to P_1.

If we work with the Principle of Non-contradiction and the Principle of Excluded Middle, we obtain the classic calculus N_0.

5.2 The N_1-Algebra

Definition 2.1 An N_1 Curry algebra (or an N_1 algebra) [3] is a classic implicative pre-lattice $[A, \equiv, \leq, \wedge, \vee, \rightarrow, ']$ with a maximum element 1, a minimum element 0 and operators $\wedge, \vee, \rightarrow$ and' satisfying the conditions below, where $x^0 =_{Def.} \left(x \wedge x'\right)'$ *and* $x^\# =_{Def.} x \vee x'$:

(1) $x^0 \wedge y^\# \leq ((x \rightarrow y) \rightarrow ((x \rightarrow y') \rightarrow x'))$
(2) $x^0 \wedge y^0 \leq (x \rightarrow y)^0 \wedge (x \wedge y)^0 \wedge (x \vee y)^0 \wedge \left(x'\right)^0$
(3) $x^\# \wedge y^\# \leq (x \rightarrow y)^\# \wedge (x \wedge y)^\# \wedge (x \vee y)^\# \wedge \left(x'\right)^\#$
(4) $x^0 \leq (x \rightarrow x'') \wedge (x \rightarrow (x' \rightarrow y))$
(5) $x^\# \leq x'' \rightarrow x$

(6) $x^0 \vee x^\# \equiv 1$.

Theorem 2.2 *Adding the postulate $(x \wedge x')' \equiv 1$we get a P_1-algebra.*

Proof Immediate.

Theorem 2.3 *Adding the postulate $(x \vee x')' \equiv 1$ we obtain a C_1-algebra.*

Proof Immediate.

Theorem 2.4 *Adding the postulates $(x \wedge x')' \equiv 1$ and $x \vee x' \equiv 1$ we obtain a Boolean pre-algebra.*

Proof Immediate.

Definition 2.5 Let x be an element of an N_1-algebra. We put $x^* =_{\text{Def.}} x' \wedge x^o$.

Theorem 2.6 *In an N_1-algebra, x^* is the Boolean complement of x; thus, $x \wedge x^* \equiv 0$ and $x \vee x^* \equiv 1$.*

Theorem 2.7 *In an N_1-algebra, the structure composed of the underlying set and the operations $\wedge$, $\vee$ and $*$ constitutes a Boolean algebra.*

Definition 2.8 Let $[A, \equiv, \leq, \wedge, \vee, \rightarrow, ']$ be an N_1-algebra and $[A, \equiv, \wedge, \vee, *]$ the Boolean algebra obtained as in the above theorem. Any Boole algebra that is isomorphic to the Boole algebra quotient of $[A, \equiv, \wedge, \vee, *]$ by $\equiv$ is called a Boolean algebra associated with the N_1-algebra $[A, \equiv, \leq, \wedge, \vee, \rightarrow, ']$.

Thus, we have the following representation theorems for N_1-algebras.

Theorem 2.9 *Every N_1-algebra is associated with a body of sets. Besides, any N_1-algebra is associated with a field of simultaneously open and closed sets of a wholly disconnected compact Hausdorff space.*

There is an open problem concerning N_1-algebras. How many non-isomorphic Boolean algebras associated with an N_1-algebra are there?

5.3 The N_n-Algebras, $1 \leq n \leq \omega$

We display a chain of Curry algebras starting with an N_1-algebra.

Let $[A, \equiv, \leq, \wedge, \vee, \rightarrow, ']$ be an N_1-algebra. If $x \in A$, remember that $x^0 =_{\text{Def.}} (x \wedge x')'$. x^1 abbreviates x^0. x^n $(1 < n < \omega)$ abbreviates $(x^{n-1})^0$. Also, $x^{(1)}$ abbreviates x^1. $x^{(n)}$ $(1 < n < \omega)$ abbreviates $x^{(n-1)} \wedge x^n$. We add that x^1 abbreviates $x^\#$. x^n $(1 < n < \omega)$ abbreviates $x^\# \wedge x^{\#\#} \wedge \ldots \wedge x^{\#\#\ldots\#}$, where the symbol # occurs n times and $x^{(n)}$ abbreviates $x^1 \wedge x^2 \wedge \ldots \wedge x^n$.

Definition 3.1 A Curry N_n algebra (or an N_n-algebra) is a classic implicative pre-lattice $[A, \equiv, \leq, \wedge, \vee, \rightarrow, ']$ ith a maximum element 1, a minimum element 0, and operators $\wedge$, $\vee$, $\rightarrow$ and' satisfying the conditions below:

(1) $x^{(n)} \wedge y^{[n]} \leq ((x \to y) \to ((x \to y') \to x'))$
(2) $x^{(n)} \wedge y^{(n)} \leq (x \to y)^{(n)} \wedge (x \wedge y)^{(n)} \wedge (x \vee y)^{(n)} \wedge \left(x'\right)^{(n)}$
(3) $x^{[n]} \wedge y^{[n]} \leq (x \to y)^{[n]} \wedge (x \wedge y)^{[n]} \wedge (x \vee y)^{[n]} \wedge \left(x'\right)^{[n]}$
(4) $x^{(n)} \leq (x \to x'') \wedge (x \to (x' \to y))$
(5) $x^{[n]} \leq x'' \to x$
(6) $x^{(n)} \vee x^{[n]} \equiv 1$.

Theorem 3.2 *Adding the postulate $x^{[n]} \equiv 1$ we obtain a C_n-algebra.*

Proof Immediate.

Theorem 3.3 *Adding the postulate $x^{(n)} \equiv 1$ we obtain a P_n-algebra.*

Proof Immediate.

Theorem 3.4 *Adding the postulates $x^{[n]} \equiv 1$ and $x^{(n)} \equiv 1$ we obtain a Boolean pre-algebra.*

Proof Immediate (Fig. 5.1).

Definition 3.5 Let x be an element of an N_n-algebra. We put $x^{*n} =_{\text{Def.}} x' \wedge x^{(n)}$.

Theorem 3.6 *In an N_n-algebra, x^{*n} is the Boolean complement of x; thus, $x \wedge x^{*n} \equiv 0$ and $x \vee x^{*n} \equiv 1$.*

Theorem 3.7 *In an N_n -algebra, the structure of the underlying set and the operations $\wedge$, $\vee$ and ^{*n} constitute a Boolean algebra.*

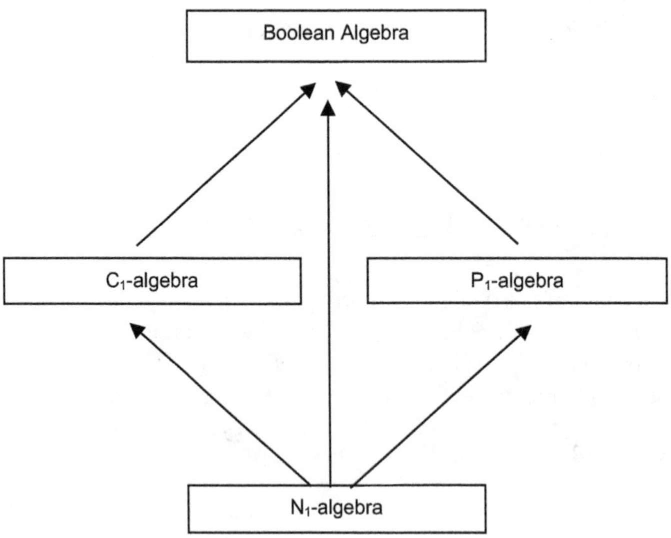

Fig. 5.1 Diagram of the correlation of C_1-algebra, P_1-algebra, N_1-algebra and Boolean algebra

Definition 3.8 Let $[A, \equiv, \leq, \wedge, \vee, \rightarrow, ']$ be an N_n-algebra and $[A, \equiv, \wedge, \vee, *^n]$ be the Boolean algebra obtained as in the above theorem. Any Boole algebra that is isomorphic to the Boole algebra quotient of $[A, \equiv, \wedge, \vee, *^n]$ by $\equiv$ is called Boole algebra associated with N_n-algebra $[A, \equiv, \leq, \wedge, \vee, \rightarrow, ']$.

Thus, we have the following representation theorems for N_n-algebras.

Theorem 3.9 *Every N_n-algebra is associated with a body of sets. Also, any N_n-algebra is associated with a body of simultaneously open and closed sets of a wholly disconnected compact Hausdorff space.*

There is an open problem concerning N_n-algebras. How many non-isomorphic Boolean algebras associated with an N_n-algebra are there?

With the necessary adaptations, it is easy for Curry algebras to extend general structural concepts such as homomorphism, monomorphism, etc.

Theorem 3.10 *Each N_n-algebra is immersed in any N_{n-1}-algebra* $(1 < n < \omega)$.

Corollary 3.10.1 Each N_n-algebra $(1 < n < \omega)$ is immersed in the N_1-algebra.

If we indicate an N_n-algebra by N_n, the hierarchy of the immersions can be represented by

$$N_1 > N_2 > \ldots > N_n > \ldots$$

5.4 The N_n-Propositional Algebras

In this session, we introduce the concept of propositional algebra N_n $(1 \leq n < \omega)$ with Rosenbloom's studies in mind [4].

Definition 4.1 A propositional algebra N_n $(1 \leq n < \omega)$ is a $[A, \tilde{A}, \rightarrow, \wedge, \vee, \neg]$ where $A \subseteq \tilde{A}$ $(A \neq \emptyset)$, $\rightarrow$, $\wedge$, and $\vee$ are binary operations and $\neg$ is a unary operator on A meeting the following conditions:

(1) $x, y \in A \Rightarrow \neg x \rightarrow (y \rightarrow x) \in \tilde{A}$
(2) $x, y, z \in A \Rightarrow (x \rightarrow y) \rightarrow ((x \rightarrow (y \rightarrow z)) \rightarrow (x \rightarrow z)) \in \tilde{A}$
(3) $(x \in A \in x \rightarrow y \in A) \Rightarrow y \in \tilde{A}$
(4) $x, y \in A \Rightarrow x \wedge y \rightarrow x \in \tilde{A}$
(5) $x, y \in A \Rightarrow x \wedge y \rightarrow y \in \tilde{A}$
(6) $x, y \in A \Rightarrow x \rightarrow (y \rightarrow x \wedge y) \in \tilde{A}$
(7) $x, y \in A \Rightarrow x \rightarrow x \wedge y \in \tilde{A}$
(8) $x, y \in A \Rightarrow y \rightarrow x \vee y \in \tilde{A}$
(9) $x, y, z \in A \Rightarrow (x \rightarrow z) \rightarrow ((y \rightarrow z)(x \vee y \rightarrow z)) \in \tilde{A}$
(10) $x, y \in A \Rightarrow x^{[n]} \wedge y^{(n)} \rightarrow ((x \rightarrow y) \rightarrow ((x \rightarrow \neg y) \rightarrow \neg x)$
(11) $x, y \in A \Rightarrow (x^{[n]} \wedge y^{(n)}) \rightarrow ((x \wedge y) \wedge ((x \vee y) \wedge (x \rightarrow y)^{(n)}$
(12) $x, y \in A \Rightarrow (x^{[n]} \wedge y^{[n]}) \rightarrow ((x \wedge y)^{[n]} \wedge (x \vee y) \wedge (x \rightarrow y)^{[n]}$

(13) $x, y \in A \Rightarrow x^{(n)} \to (x \to \neg\neg x) \wedge (x \to (\neg x \to y))$
(14) $x \in A \Rightarrow x^{[n]} \to (\neg\neg x \to x)$
(15) $x \in A \Rightarrow x^{(n)} \vee x^{[n]} \equiv 1.$

Definition 4.2

(1) $x \sim y$ abbreviates $(x \to y) \wedge (y \to x)$
(2) $\vdash x$ abbreviates $x \in \tilde{A}$

Theorem 4.3 *In a propositional algebra N_1, we have:*

(1) $\vdash x \wedge x \to y$
(2) $\vdash x \vee (x \to y)$
(3) $\vdash x^{\#} \to (\neg\neg x \leftrightarrow x)$
(4) $\vdash x^{\#} \to ((x \to y) \wedge (x \to \neg y) \to x)$
(5) $\vdash x \wedge \neg x \leftrightarrow y \wedge \neg y$

Theorem 4.4 *In a propositional algebra N_1, in general, we do not have:*

(1) $\vdash (x \wedge \neg x) \to y$;
(2) $\vdash (x \wedge \neg x) \to \neg y$;
(3) $\vdash (x \to (\neg x \to y))$;
(4) $\vdash (x \leftrightarrow \neg x) \to y$;
(5) $\vdash \neg\neg x \to x$;
(6) $\vdash (x \leftrightarrow \neg x) \to \neg y$;
(7) $\vdash x \to \neg\neg x$;
(8) $\vdash (\neg x \wedge (x \vee y)) \to y$;
(9) $(x \to y) \to (\neg y \to \neg x)$;
(10) $\vdash \neg(x \vee y) \leftrightarrow (\neg x \wedge \neg y)$;
(11) $\vdash \neg(x \wedge y) \leftrightarrow (\neg x \vee \neg y)$;
(12) $\vdash (x \wedge \neg x) \leftrightarrow (y \wedge \neg y)$;

Theorem 4.5 *Adding the postulate $\neg(x \wedge \neg x) \equiv 1$ we get a propositional algebra P_1.*

Proof Immediate.

Theorem 4.6 *Adding the postulate $x \vee \neg x \equiv 1$. we obtain a propositional algebra C_1.*

Proof Immediate (Fig. 5.2 below).

Theorem 4.7 *Adding the postulates $\neg(x \wedge \neg x) \equiv 1$ and $x \vee \neg x \equiv 1$. we obtain a propositional Boolean algebra.*

Proof Immediate.

Definition 4.8 Let x be an element of an N_1-algebra. We put $x^* =_{\text{Def.}} \neg x \wedge x^{o}$.

Theorem 4.9 *In an N_1-algebra, x^* is the Boolean complement of x; thus $x \wedge x^* \equiv 0$ and $x \vee x^* \equiv 1$.*

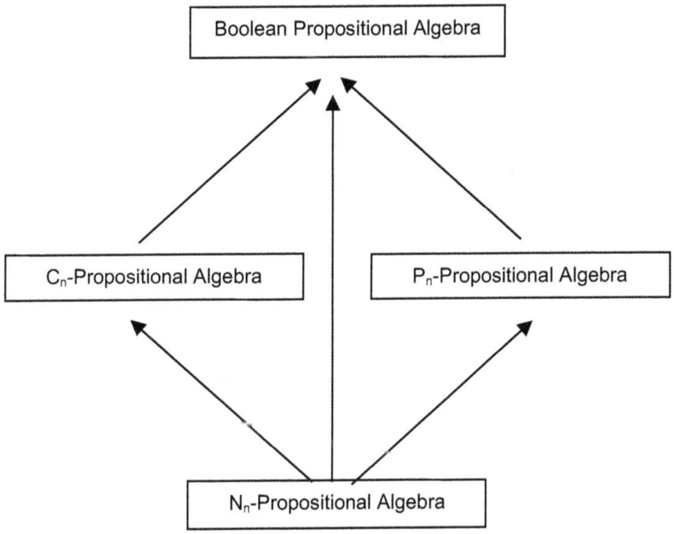

Fig. 5.2 Diagram of the correlations of the propositional algebra C_n-propositional algebra, P_n-propositional algebra, N_n-propositional algebra, and Boolean propositional algebra

Theorem 4.10 *In a propositional algebra N_1, in general, we do not have:*

(1) $\vdash x \vee \neg x$
(2) $\vdash \neg(x \vee y) \leftrightarrow \neg x \wedge \neg y$
(3) $\vdash \neg(x \vee y) \leftrightarrow \neg x \wedge \neg y$
(4) $\vdash \neg\neg x \to x$
(5) $\vdash \neg\neg x \leftrightarrow x$
(6) $\vdash (x \to y) \to (\neg y \to \neg x)$
(7) $\vdash x^{\#\#}$.

Proof Immediate.

Definition 4.11

(1) $x \equiv y$ abbreviates $\vdash x \leftrightarrow y$ and $x \leq y$ abbreviates $\vdash x \to y$.

We have $x \equiv y \Leftrightarrow x \leq y$ and $y \leq x$.

Theorem 4.12 *We have*:

(1) $\equiv$ is an equivalence relation
(2) $\leq$ is a quasi-order
(3) $x \leq y \Leftrightarrow x \wedge y \equiv x$
(4) $x \leq y \Leftrightarrow x \vee y \equiv y$
(5) $0 \leq x \leq 1$.

Theorem 4.13 *If* $[A, A, \rightarrow, \wedge, \vee, \neg]$ *is a propositional algebra* N_1*, if we define* $\equiv$ *and* $0 =_{Def} x^0 \wedge \neg x^0$ *and if' abbreviates* $\neg$*, then* $[A, \equiv, \rightarrow, \wedge, \vee, 0, ']$ *is a Curry* N_1 *algebra.*

Theorem 4.14 *If* $[A, \equiv, \rightarrow, \wedge, \vee, 0, ']$ *is a propositional algebra* N_1*, if we define* $\tilde{A}$ *as the set of elements x such that* $x \equiv y \rightarrow y$ *for some* $y \in A$*, and represent the operator' by* $\neg$*, the structure* $[A, \tilde{A}, \rightarrow, \wedge, \vee, \neg]$ *is a propositional algebra* N_1.

Theorem 4.15 *The concepts of Curry algebra* N_1 *and propositional algebra* N_1 *are equivalent (in the sense of Bourbaki).*

Theorem 4.16 *In a propositional algebra* N_1, $[A, \tilde{A}, \rightarrow, \wedge, \vee, \neg]$ *the operator* $\neg$ *is not monotonic concerning* $\equiv$.

Proof Adaptation of the Proof made in [5], remembering that propositional algebra N_1 has some properties of propositional algebra C_1.

At this stage, we can introduce the usual studies in algebraic structures, like homomorphism, representation theorems, or applications for the study of completeness of N_1 calculus, and so on. As the subject is analogous to the study made with C_1-algebras, we will omit such details. So the interested readers can complete the topics themselves as an exercise.

References

1. da Costa, N.C.A.: On the theory of inconsistent formal systems. Notre Dame, J. Formal Log. **15**, 497–510 (1974). https://doi.org/10.1305/ndjfl/1093891487
2. da Costa, N.C.A.: Logics that are both paraconsistent and paracomplete. Atti della Accademia Nazionale dei Lincei. Classe di Scienze Fisiche, Matematiche e Naturali. Rendiconti Lincei. Mat. e Appl. **83**(1), 29–32 (1989). http://eudml.org/doc/287311
3. Abe, J.M.: Curry algebras N_1, Atti della Accademia Nazionale dei Lincei. Classe di Scienze Fisiche, Matematiche e Naturali. Rendiconti Lincei. Mat. e Appl., Ser. 9, **7** (3), 125–128 (1996), ISSN: 1120–6330
4. Sikorski, R.: Boolean algebras. Springer-Verlag, (1969). ISBN-13: 978-3540044697, ISBN-10: 3540044698
5. Grätzer, G.: Universal algebra, 2nd ed, p. 583. Springer-Verlag, (1979). ISBN-10: 0387903550; ISBN-13: 978–0387903552

Chapter 6
Algebraization of 1st Order Paraconsistent Systems C_n* ($1 \leq n < \omega$)

In this chapter, we will be engaged in the algebraization of da Costa 1st order C_n^* ($1 \leq n < \omega$) systems.

6.1 The Systems C_n* ($1 \leq n < \omega$)

As discussed in paragraph 1, Chap. 2, we present a family of paraconsistent first-order predicate calculus C_n^* ($1 \leq n < \omega$) introduced by da Costa [1]. For each n, ($1 \leq n < \omega$), we have a different calculus symbolized by C^*_n. Such calculus was formulated in order to satisfy the following conditions:

(1) The Principle of Non-contradiction, in the form $\neg(A \wedge \neg A)$, is not generally valid;
(2) We cannot deduce any formula B from two contradictory propositions, A and $\neg A$.
(3) The calculus must contain the classical propositional calculus's essential schemes and inference rules compatible with conditions (a) and (b) above.

The language of the calculus C_n^* ($1 \leq n < \omega$) is the same for all of them; let us denote by L. The primitive symbols of L are the following:

(1) Individual variables: x, y, z…
(2) For each n, n-ary functional symbols. The 0-ary functional symbols are also called individual constants
(3) For each n, symbols of n-ary predicates
(4) The equality symbol $=$
(5) The symbols $\neg$, $\wedge$, $\vee$, $\rightarrow$, $\exists$ and $\forall$
(6) Auxiliary symbols (,) and.

General syntactic concepts are introduced, such as the idea of a formula. Let A be a formula.

J. M. Abe, *An Introduction to Curry Systems*, Intelligent Systems Reference Library 280, https://doi.org/10.1007/978-3-031-97810-4_6

Definition 1.1 Let A be a formula. We have:

A^o abbreviates $\neg(A \wedge \neg A)$.
A^1 abbreviates A^o,
A^{n+1} abbreviates $(A^n)^o$, $n \in N$.

Definition 1.2 Let A be a formula. We have:

$A^{(1)}$ abbreviates $(A \wedge A^1)$.
$A^{(n+1)}$ abbreviates $A^{(n)} \wedge A^{n+1}$, $n \in N$.

The postulates (axiom schemes and inference rules) of C_n* ($1 \leq n < \omega$) are as follows: A, B, and C denote any formulas.

$(\rightarrow_1) A \rightarrow (B \rightarrow A)$

$(\rightarrow_2) (A \rightarrow (B \rightarrow C)) \rightarrow ((A \rightarrow B) \rightarrow (A \rightarrow C))$

$(\rightarrow_3) \dfrac{A, A \rightarrow B}{B}$ (MP (Modus Ponens))

$(\wedge_1) A \wedge B \rightarrow A$

$(\wedge_2) A \wedge B \rightarrow B$

$(\wedge_3) A \rightarrow (B \rightarrow (A \wedge B))$

$(\vee_1) A \rightarrow A \vee B$

$(\vee_2) B \rightarrow A \vee B$

$(\vee_3)\ (A \rightarrow C) \rightarrow ((B \rightarrow C) \rightarrow ((A \vee B) \rightarrow C))$

$(\neg_1) B^{(n)} \rightarrow ((A \rightarrow B) \rightarrow ((A \rightarrow \neg B) \rightarrow \neg A)$

$(\neg_2)\ (A^{(n)} \wedge B^{(n)}) \rightarrow ((A \wedge B)^{(n)} \wedge (A \vee B)^{(n)} \wedge (A \rightarrow B)^{(n)})$

$(\neg_3)\ (A \vee \neg A)$

$(\neg_4) \neg\neg A \rightarrow A$

$(\forall_1) \forall x A(x) \rightarrow A(t)$

$(\forall_2) \dfrac{B \rightarrow A(x)}{B \rightarrow \forall x A(x)}$

$(\exists_1) A(t) \rightarrow \exists x A(x)$

$(\exists_2) \dfrac{A(x) \rightarrow B}{\exists x A(x) \rightarrow B}$

$(\forall_3) \forall x (A(x))^{(n)} \rightarrow (\forall x A(x))^{(n)}$

$(\exists_3) \forall x (A(x))^{(n)} \rightarrow (\exists x A(x))^{(n)}$

with the usual restrictions

(C) If A and B are congruent formulas (in the sense of Kleene [2]), or if one results from the other through the suppression of vacuous quantifiers, then $A \leftrightarrow B$ constitutes an axiom.

We write $\neg^{(i)} A$ for $\neg A \wedge A^{(i)}$, i = 1, 2,…, n.

We have the following result correlating to classical positive logic: in C_n^* $(1 \leq n < \omega)$, all valid schemes of classical positive propositional logic apply. In particular, the deduction theorem is valid in C_n^* $(1 \leq n < \omega)$. C_ω* contains positive intuitionist logic.

In C_n^* $(1 \leq n < \omega)$, we have the following meta-theorem:

$$\vdash \neg A^{(n)} \rightarrow (\neg A)^{(n)}$$

The following comments mainly apply to C_n* calculus, as they are an extension of C_n calculus. In calculus C_n^* $(1 \leq n < \omega)$ formula $A^{(i)}$ intuitively expresses that formula A behaves classically; therefore, the motivation for the postulates $(\neg_3)$ and $(\neg_4)$ is clear. Furthermore, the postulates show us that the other connectors $\wedge$, $\vee$, $\rightarrow$ have all the properties of the classic conjunction, disjunction, and implication. Also, the defined connective $\neg^{(i)}$ has all the properties of classical negation. Consequently, the classical propositional calculus is contained in C_n^* $(1 \leq n < \omega)$ although the latter constitutes a strict under-calculus of the former. It is immediate to verify that conditions (a), (b), and (c) above are satisfied. Thus, the Principle of Non-contradiction $\neg(A \wedge \neg A)$ is generally not valid.

One can introduce a semantics for such a calculus, known as valuation semantics. Let A and B be any formulas. F symbolizes the set of formulas for C_1, and 2 indicates the set {0, 1}. An interpretation (or validation) of C_1 is a function φ: $F \rightarrow 2$ such that:

(1) $\varphi(A) = 0 \Rightarrow \varphi(\neg A) = 1$;
(2) $\varphi(\neg\neg A) = 1 \Rightarrow \varphi(A) = 1$;
(3) $\varphi(B^o) = \varphi(A \rightarrow B) = \varphi(A \rightarrow \neg B) = 1 \Rightarrow \varphi(A) = 0$;
(4) $\varphi(A \rightarrow B) = 0 \Rightarrow \varphi(A) = 0$ or $\varphi(B) = 1$;
(5) $\varphi(A \wedge B) = 1 \Rightarrow \varphi(A) = \varphi(B) = 1$;
(6) $\varphi(A \vee B) = 1 \Rightarrow \varphi(A) = 1$ or $\varphi(B) = 1$;
(7) $\varphi(A^o) = \varphi(B^o) = 1 \Rightarrow \varphi((A \rightarrow B)^o) = \varphi((A \wedge B)^o) = \varphi((A \vee B)^o) = 1$.

6.2 The C_1*-Monadic Algebras

This section introduces the monadic Curry algebra C_1* [3].

Definition 2.1 Let A be a C_1-algebra. Let ∃ (existential quantifier) and ∀ (universal quantifier) be operators on A. (∃, ∀) is called a quantifier on A if.

(1) $\exists 0 \equiv 0$
(2) $x \leq \exists x$
(3) $\exists(x \vee \exists y) \equiv \exists x \vee \exists y$
(4) $\exists\exists x \equiv \exists x$
(5) $\exists(\exists x)^* \equiv (\exists x)^*$
(6) $\exists(x \wedge \exists y) \equiv \exists x \wedge \exists y$

(7) $\forall 1 \equiv 1$
(8) $\forall x \leq x$
(9) $\forall(x \vee y) \equiv \forall x \wedge \forall y$
(10) $\forall\forall x \equiv \forall x$
(11) $\forall(\forall x)^* \equiv (\forall x)^*$.

We suppose in the above definition that, if $x \equiv y$, then $\exists x \equiv \exists y$ and $\forall x \equiv \forall y$. $\exists$ is called existential quantifier on A, and $\forall$ is called universal quantifier on A. The pair $< A, (\exists, \forall) >$ is called a monadic Curry algebra C_1* or a C_1*-monadic algebra (or C_1*-algebra).

Given a Curry algebra C_1, let us assume that there is a universal quantifier defined on it, i.e., a structure $< A, \forall >$ such that conditions 7–11 above are satisfied. If we define $\exists_1 x =_{\text{Def.}} (\forall x^*)^*$, then $\exists_1$ is an existential quantifier (i.e., satisfying 1–6), and the structure composed by $[A, (\exists_1, \forall)]$ is a C_1*- monadic algebra. Also, we can get a monadic algebra considering an existential quantifier $\exists$ on a Curry algebra C_1 satisfying conditions 1–6 of the above definition and defining a universal quantifier (i.e., satisfying 7–11) as $\forall_1 x =_{\text{Def.}} (\exists x^*)^*$. Then the structure composed by $[A, (\exists, \forall_1)]$ is a C_1*- monadic algebra. Given a Curry algebra C_1, the algebras obtained $[A, (\exists_1, \forall)]$ and $[A, (\exists, \forall_1)]$ are not in general isomorphic. Also, given a C_1*- monadic algebra $[A, (\exists, \forall)]$, define the new quantifiers $\exists_1$ and $\forall_1$. In general, we have $\exists_1 \neq \exists$ and $\forall_1 \neq \forall$.

Let C be a C_1-algebra and $A = \{(x_1, x_2, \ldots, x_n) \mid x_i \in C, i = 1, 2, \ldots, n\}$. Let us suppose that if $x, y \in A$, then $x \vee y \in A$ and $\neg x \in A$, and if $(x_1, x_2, \ldots, x_n), (y_1, y_2, \ldots, y_n) \in A$ we define $(x_1, x_2, \ldots, x_n) \equiv (y_1, y_2, \ldots, y_n)$ iff $x_i \equiv y_i$, $i = 1, 2, \ldots, n$. Also, we put $\overline{x} = \bigvee_{i=1}^{n} x_i$, $\underline{x} = \bigwedge_{i=1}^{n} x_i$, and assume that $(\underbrace{\overline{x}, \ldots, \overline{x}}_{n-times}), (\underbrace{\underline{x}, \ldots, \underline{x}}_{n-times}) \in A$. If we define $\exists x = (\overline{x}, \ldots, \overline{x})$ and $\forall x = (\underline{x}, \ldots, \underline{x})$, then A is a C_1*-monadic algebra.

A more useful example of C_1*-monadic algebra is the following. Let C be a C_1-algebra, a set $K \neq \emptyset$, and C^K is the set of all functions of K into C. Let A be the set such that: (i) A is a C_1-algebra concerning the pointwise operations, and (ii) if $x \in A$, then the range of x has a supremum $\overline{x}$ and an infimum $\underline{x}$ in C, and the functions that take the value $\overline{x}$ at each point of K and $\underline{x}$ at each point of K are in A. If $\exists x$ and $\forall x$ are defined to be those functions, then A becomes a C_1*-monadic algebra. Every C_1*-monadic algebra obtained in this way is called a C-valued functional algebra with domain K [4].

6.3 C_1*-Monadic Algebras and Monadic Algebras

Now, we discuss some aspects of the algebraic structures that originated in our discussion. The matter is very rich, but due to this paragraph's limitations, we will be concerned only with some of them.

Theorem 3.1 *In a C_1*-monadic algebra [A, (∃, ∀)], the structure composed by the underlying set and by operations ∧, ∨, *, ∃, and ∀ is a (pre) monadic algebra. If we pass to the quotient by the fundamental relation ≡, we obtain a monadic algebra in the usual sense* [4].

Definition 3.2 Let $[A, (\exists, \forall)]$ be a C_1*-monadic algebra, and $[A, \equiv, \leq, \rightarrow, {}^*, \exists, \forall]$ the monadic algebra obtained as in the above theorem. Any monadic algebra that is isomorphic to the quotient algebra of $[A, \equiv, \leq, \rightarrow, {}^*, \exists, \forall]$ by ≡ is called monadic algebra associated with the C_1*-monadic algebra.

Hence, we can establish the following representation theorems for C_1*-monadic algebras.

Theorem 3.3 *If C is a C_1*-monadic algebra, then for its associated monadic algebra A, there exists a set X, and there exits a Boolean algebra B, such that (i) A is isomorphic to a B-valued functional algebra A' with domain X, and (ii) for every element p of A' there exists a point x in X with $p(x) = \exists p(x)$.*

Theorems 3.1 and 3.2 show us that C_1*-monadic algebras constitute an interesting generalization of the concept of monadic algebras. Here, there is an open problem. How many non-isomorphic monadic algebras are associated with a C_1*-monadic algebra?

References

1. da Costa, N.C.A.: On the theory of inconsistent formal systems. Notre Dame, J. Formal Log. **15**, 497–510 (1974). https://doi.org/10.1305/ndjfl/1093891487
2. Kleene, S.C.: Introduction to metamathematics, Ishi Press, Japan (2009). ISBN-10: 0923891579, ISBN-13: 978-0923891572, p. 572
3. Abe, J.M., Nakamatsu, K., Akama, S.: An algebraic version of the monadic system C_1, new advances in intelligent decision technologies, series: Studies in computational intelligence, vol. 199. In: Nakamatsu, K., Phillips-Wren, G.. Jain, L.C., Howlett, R.J. (eds.), XVI, 656 p. 234 illus., Hardcover, Springer Berlin/Heidelberg, (2009). ISBN: 978–3–642–00908–2, 341–349. https://doi.org/10.1007/978-3-642-00909-9
4. Halmos, P.H.: Algebraic logic, Dover Publications (September 12, 2018). ISBN-10: 0486828042, ISBN-13: 978-0486828046
5. Curry, H.B.: Leçons de Logique Algébrique, Gauthiers-Villars, p. 163 (1952)
6. Fidel, M.: The decidability of the calculi C_n. Rep. Math. Log. **8**, 31–40
7. Abe, J.M.: Curry algebras Pτ. Log. Anal., Nouv. Le Série, **41** (161/163), 5–15 (1998). Peeters Publishers. https://doi.org/10.2307/44084618. https://www.jstor.org/stable/44084618

Chapter 7
Algebraization of 1st Order Paracomplete Systems P_n* ($1 \leq n < \omega$)

7.1 Introduction

In one of its possible formulations, the Principle of Excluded Middle says that one is true of two contradictory propositions, A and $\neg A$ (the negation of A). A paracomplete logic is a logic in which a proposition and its negation can be false. Therefore, paracomplete logic derogates the Principle of Excluded Middle. More precisely, a logic is said to be complete if there are non-trivial maximal sets such that a proposition and its negation do not belong to them. In this sense, many-valued logic comprises several types, such as intuitionist logic and several many-valued logics.

A motivation for paracomplete logic is that the classic condition "of two contradictory propositions, one is true" does not always reflect our intuitions. For example, if P is a vague predicate and x is a borderline individual of vagueness, we can assume that sometimes P(x) and ¬P(x) are false. This happens because we know that most concepts related to the real world have a certain degree of imprecision, naturally leading to the need for logical systems with characteristics of paracomplete logic. They are also related to ideas that underlie fuzzy logic.

7.2 The P_n* ($1 \leq n < \omega$) Calculus

In [1], da Costa and Marconi built a P_n ($1 \leq n < w$), of paracomplete propositional calculus. These calculi are 'dual' of the paraconsistent propositional logic, C_n ($1 \leq n < w$), In the same reference, the authors note that the $\mathrm{P_n}$ calculus ($1 \leq n < w$) can be extended to first-order calculus P^*_{n} ($1 \leq n <$ w), which we will outline in this chapter.

For each n, $1 \leq n <$ w, we have different calculus, represented by P^*_n. Such calculus was formulated in order to satisfy the following conditions:

J. M. Abe, *An Introduction to Curry Systems*, Intelligent Systems Reference Library 280, https://doi.org/10.1007/978-3-031-97810-4_7

(1) The Principle of Excluded Middle, in the form $A \wedge \neg A$ is not generally valid;
(2) There are formulas A and ¬A, which are both false.
(3) The calculus must contain the classical propositional calculus's essential schemes and inference rules compatible with conditions a) and b) above.

The language of the P_n^* $(1 \leq n < w)$ calculus is the same for all of them; Let us denote by L. The language L is a first-order language whose primitive symbols are the following:

(1) Individual variables: an infinite denumerable set of symbols.
(2) Logical connectives: ¬ (negation), ∧ (conjunction), ∨ (disjunction), and → (implication).
(3) For each n, zero or more n-ary function symbols (n is a natural number).
(4) For each n, zero or more n-ary predicate symbols.
(5) Quantifiers: ∀ (universal) and ∃ (existential).
(6) Auxiliary symbols: parentheses and commas.

For each n, the number of function symbols can be zero or not, finite or infinite. A 0-ary function symbol is called a constant. Also, the number of predicate symbols can be finite or infinite.

Formulas are defined in the usual way.

Definition 2.1 Let A be a formula. We have that $A^{\#}$ abbreviates $A \vee \neg A$. We also $A^{[\mathrm{n}]}$ for $A^{\#} \vee A^{\#\#} \vee A^{\#\#\#} \vee \ldots \vee A^{\#\mathrm{n}}$, where #n means that the symbol # appears n times.

Definition 2.2 We write $(A \leftrightarrow B)$ for $(A \rightarrow B) \wedge (B \rightarrow A)$.

They are postulates (axiom schemes and primitive rules) of P_n^*.

$(\rightarrow_1)$ $A \rightarrow (B \rightarrow A)$

$(\rightarrow_2)$ $(A \rightarrow (B \rightarrow C)) \rightarrow ((A \rightarrow B) \rightarrow (A \rightarrow C))$
$(\rightarrow_3)$ $\frac{A, A \rightarrow B}{B}$ (MP (Modus Ponens))
$(\rightarrow_4)$ $((A \rightarrow B) \rightarrow A) \rightarrow A$(Peirce's law)
$(\wedge_1)$ $A \wedge B \rightarrow A$
$(\wedge_2)$ $A \wedge B \rightarrow B$
$(\wedge_3)$ $A \rightarrow (B \rightarrow (A \wedge B))$
$(\vee_1)$ $A \rightarrow A \vee B$
$(\vee_2)$ $B \rightarrow A \vee B$
$(\vee_3)$ $(A \rightarrow C) \rightarrow ((B \rightarrow C) \rightarrow ((A \vee B) \rightarrow C))$
$(\neg_5)$ $\neg(\neg A \wedge \neg A)$
$(\neg_6)$ $A \rightarrow (\neg A \rightarrow B)$
$(\neg_7)$ $A \rightarrow \neg\neg A$
$(\neg_8)$ $A^{[\mathrm{n}]} \rightarrow ((A \rightarrow B) \rightarrow ((A \rightarrow \neg B) \rightarrow \neg A))$
$(\neg_9)$ $(A^{[n]} \wedge B^{[n]}) \rightarrow ((A \wedge B)^{[n]} \wedge (A \vee B)^{[n]} \wedge (A \rightarrow B)^{[n]})$
$(\exists_1)$ $A(t) \rightarrow \exists x A(x)$
$(\exists_2)$ $\frac{A(x) \rightarrow B}{\exists x A(x) \rightarrow B}$
$(\forall_1)$ $\forall x A(x) \rightarrow A(t)$

($\forall_2$) $\frac{B \rightarrow A(x)}{B \rightarrow \forall x A(x)}$

($\forall_4$) $\forall x(A(x))^{[n]} \rightarrow (\forall x A(x))^{[n]}$

($\exists_4$) $\forall x(A(x))^{[n]} \rightarrow (\exists x A(x))^{[n]}$

with the usual restrictions.

(C) If A and B are congruent formulas [2] or if one results from another by suppressing vacuous quantifiers, then $A \leftrightarrow B$ constitutes an axiom.

The calculus P^*_ω is obtained from P_n by suppressing the axiom schemes ($\neg 8$) and ($\neg 9$) with the addition of the schemes ($\exists_1$), ($\exists_2$), ($\forall_1$), ($\forall_2$)*and*(C).

Theorem 2.3 *In P^*_n ($1 \leq n < \omega$), all the theorems and rules of the classical positive calculus are valid.*

Proof It is immediate, considering the postulates of P^*_n ($1 \leq n < \omega$).

Theorem 2.4 *In P^*_n ($1 \leq n < \omega$), the following schemes are valid:*

(1) $\vdash (A \wedge \neg A) \rightarrow B$
(2) $\vdash A \vee (A \rightarrow B)$
(3) $\vdash A^{\#} \rightarrow (\neg\neg A \leftrightarrow A)$
(4) $\vdash A^{\#} \rightarrow ((A \rightarrow B) \rightarrow (\neg A))$
(5) $\vdash (A \wedge \neg A) \leftrightarrow (B \wedge \neg B)$.

Proof Immediate.

Theorem 2.5 *In P^*_n ($1 \leq n < w$), the schemes are not valid:*

(1) $A \wedge \neg A$
(2) $\neg\neg A \leftrightarrow A$
(3) $\neg(A \wedge B) \leftrightarrow \neg A \vee \neg B$
(4) $\neg(A \vee B) \leftrightarrow \neg A \wedge \neg B$
(5) $(A \rightarrow B) \rightarrow (\neg B \rightarrow \neg A)$
(6) $A^{\#\#}$.

Proof Immediate.

Definition 2.6 In P^*_n ($1 \leq n < \omega$), we define a "strong negation", $\neg^*$, as follows: $\neg^* A =_{Def.} A \rightarrow (B \wedge \neg B)$ (where B is a fixed formula).

Theorem 2.7 *In P^*_n ($1 \leq n < w$) the operator strong negation $\neg^*$ together with the remaining connectives $\wedge$, $\vee$, and $\rightarrow$ they have all properties of classical connectives of negation, conjunction, disjunction, and implication, respectively.*

Proof Immediate.

Corollary 2.7.1 In $P^*_n(1 \leq n < w)$ the schemes are valid:

(1) $((A \rightarrow B) \rightarrow ((A \rightarrow \neg^{\#} B) \rightarrow \neg^{\#} A)$

(2) $A \rightarrow (\neg^{\#} A \rightarrow B)$
(3) $A \vee \neg^{\#} A$.

Proof Due to the previous theorem.

Theorem 2.8 *If P_0^* denotes the classical predicate calculus, then each sequence calculus, starting from the second, $P_0^*, P_1^*, P_2^*, \ldots, P_n^*, \ldots, P_\omega^*$ is strictly stronger than each preceding.*

Proof Analogous to the case of C_n* calculus.

7.3 P_1*-Monadic Algebras

This section presents some monadic Curry algebras P_1^* [3].

Such algebraic systems constitute one of the possible versions of the monadic calculus P_1* seen earlier.

Definition 3.1 Let A be a P_1-algebra. Let $\exists$ and $\forall$ be operators on A. $(\exists, \forall)$ is called a quantifier on A if:

(1) $\exists 0 \equiv 0$;
(2) $x \leq \exists x$;
(3) $\exists(x \vee y) \equiv \exists x \vee \exists y$;
(4) $\exists\exists x \equiv \exists x$;
(5) $\exists(\exists x)^* \equiv (\exists x)^*$;
(6) $\exists(x \wedge \exists y) \equiv \exists x \wedge \exists y$;
(7) $\forall 1 \equiv 1$;
(8) $\forall x \leq x$;
(9) $\forall(x \vee y) \equiv \forall x \vee \forall y$;
(10) $\forall\forall x \equiv \forall x$;
(11) $\forall(\forall x)^* \equiv (\forall x)^*$.

We suppose in the above definition that, if $x \equiv y$, then $\exists x \equiv \exists y$ and $\forall x \equiv \forall y$.$\exists$ is called existential quantifier on A and $\forall$ is called universal quantifier on A. The pair $[A, (\exists, \forall)]$ is called a monadic Curry algebra P_1^* or a P_1^*-monadic algebra (or P_1^*−algebra).

Given a Curry algebra P_1, let's assume that there is a universal quantifier defined on it, i.e., a structure $(A, \forall)$ such that conditions (7)–(11) above are satisfied. If we define $\exists_1 x =_{Def.} (\forall x^*)^*$, then $\exists_1$ is an existential quantifier (i.e., satisfying (1)–(6), and the structure composed by $[A, (\exists_1, \forall)]$ $[A, (\exists_1, \forall)]$ is a P_1^*− P_1^*− monadic algebra.

Also, we can get a monadic algebra considering an existential quantifier $\exists$ on a Curry algebra P_1 satisfying conditions (1)–(6) of the above definition and defining a

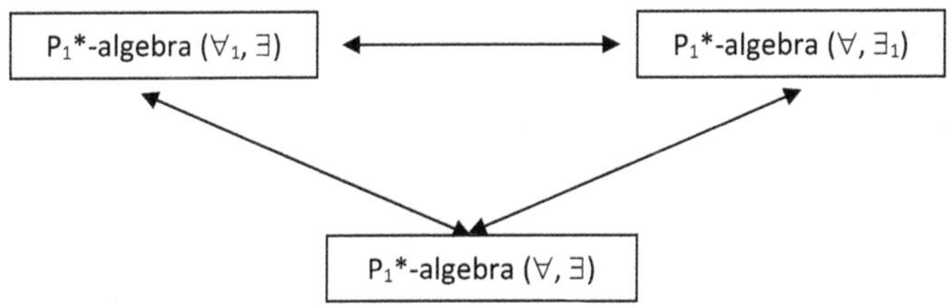

Fig. 7.1 Obtaining P_1∗-monadic algebras

universal quantifier (i.e., satisfying (7)–(11) as $\forall_1 x =_{Def.} (\exists x^*)^*$. Then the structure composed by [A, $(\exists, \forall_1)$] is a P_1*-monadic algebra.

Given a Curry algebra P_1 the algebras obtained $[A, (\exists_1, \forall)]$ and $[A, (\exists, \forall_1)]$ are not in general isomorphic.

Also, given a P_1^*-monadic algebra $[A, (\exists, \forall)]$, define the new quantifiers $\exists_1$ and $\forall_1$. In general, we have $\exists_1 \neq \exists$ and $\forall_1 \neq \forall$ (Fig. 7.1).

Let P be a P_1-algebra and $A = \{(x_1, x_2, \ldots, x_n) | x_i \in P, i = 1, 2, \ldots, n\}$. Let us suppose that if $x, y \in A$, then $x \vee y \in A$ and $x' \in A$, and if $(x_1, x_2, \ldots, x_n), (y_1, y_2, \ldots, y_n) \in A$ we define $(x_1, x_2, \ldots, x_n) \equiv (y_1, y_2, \ldots, y_n)$ iff $x_i \equiv y_i, i = 1, 2, \ldots, n$. Also, we put $\overline{x} = \bigvee_{i=1}^{n} x_i, \underline{x} = \bigwedge_{i-1}^{n} x_i$, and assume that $\underbrace{(\overline{x}, \ldots, \overline{x})}_{n-times}, \underbrace{(\underline{x}, \ldots, \underline{x})}_{n-times} \in A$ If we define $\exists x = (\overline{x}, \ldots, \overline{x})$ and $\forall x = (\underline{x}, \ldots, \underline{x})$, then A is a P_1^*-monadic algebra.

A more useful example of P_1^*-monadic algebra is the following. Let P be a P_1-algebra, a set $K^1 \neq \emptyset$, and P^K is the set of all functions of K into P. Let A be the set such that: (i) A is a P_1-algebra concerning the pointwise operations, and (ii) if $x \in A$, then the range of x has a supremum $\overline{x}$ and an infimum $\underline{x}$ in P, and the functions that take the value $\overline{x}$ at each point of K and $\underline{x}$ each point of K are in A. If $\exists x$ and $\forall x$ are defined as those functions, then A becomes a P_1^*-monadic algebra. Every P_1^*-monadic algebra obtained in this way is called a P-valued functional algebra with domain K.

7.4 P_1*-Monadic Algebras and Monadic Algebras

Now, we will discuss some of the results of the algebraic structures that originated in our discussion. The matter is very rich, but due to this paragraph's limitations, we will only be concerned with some of them.

Theorem 4.1 *In a* P_1**-monadic algebra* $[A, (\exists, \forall)]$, *the structure composed by the underlying set and by operations* $\wedge, \vee, {}^*, \exists$, *and* $\forall$ *is a (pre) monadic algebra. If we pass to the quotient by the fundamental relation* $\equiv$, *we obtain a monadic algebra in the usual sense* [4].

Definition 4.2 Let $[A, (\exists, \forall)]$ be a P_1*-monadic algebra, and $[A, \equiv, \leq, \rightarrow, {}^*, \exists, \forall]$ the monadic algebra obtained as in the above theorem. Any monadic algebra that is isomorphic to the quotient algebra of $[A, \equiv, \leq, \rightarrow, {}^*, \exists, \forall]$ by $\equiv$ is called a monadic algebra associated with the P_1*-monadic algebra.

Hence, we can establish the following representation theorems for P_1*-monadic algebras.

Theorem 4.3 *If P is a P_1*-monadic algebra, then for its associated monadic algebra A, there exists a set X, and there exists a Boolean algebra B, such that.*

(1) *A* is isomorphic to a B-valued functional algebra *A*' with domain X, and
(2) For every element *p* of *A*', there exists a point x in X such that $p(x) = \exists p(x)$.

Theorems 4.2 and 4.3 show us that P_1*-monadic algebras constitute an interesting generalization of monadic algebras. Here, there is an open problem. How many non-isomorphic monadic algebras are associated with a P_1*-monadic algebra?

Another remaining work is to study the monadic systems corresponding to P_n-algebras $(1 < n < \omega)$.

References

1. da Costa, Newton C.A., Marconi, D.: A note on paracomplete logic. Atti della Accademia Nazionale dei Lincei. Classe di Scienze Fisiche, Matematiche e Naturali. Rendiconti Lincei. Matematica e Applicazioni **80**(7–12), 504–509 (1986)
2. Kleene, S.C.: Introduction to metamathematics, Ishi Press, Japan. ISBN-10: 0923891579, ISBN-13: 978-0923891572, p. 572 (2009)
3. Abe, J.M., Nakamatsu, K., Akama, S.: A note on monadic curry system P₁, Lecture notes in computer science—LNAI 5712, pp. 388–394. ISSN 0302–9743, Springer-Verlag, Berlin/Heidelberg, Alemanha, (2009). https://doi.org/10.1007/978-3-642-04592-9_49
4. Halmos, P.H.: Algebraic logic, Dover Publications (September 12, 2018). ISBN-10: 0486828042, ISBN-13: 978-0486828046
5. Barros, C.M., Da Costa, N.C.A., Abe, J.M.: Tópico de Teoria dos Sistemas Ordenados, Vol II: Sistemas de Curry, Coleção Documentos, Série Lógica e Teoria da Ciência, IEA-USP, (20), p. 132 (1995). ISSN 16799429

Chapter 8
Algebraization of 1st Order Non-Alethic Systems N_n* (1 ≤ n < ω)

8.1 Introduction

Non-alethic logics were discovered when the duality of C_n^* $(1 \leq n < w)$ paraconsistent systems were clarified. Such dual systems are the corresponding paracomplete P_n^* $(1 \leq n < w)$ that violate the Principle of Excluded Middle. It was found that logics with paraconsistent and paracomplete characteristics could be built simultaneously. Such logics have been named non-alethic. In [1], N_n^* $(1 \leq n < w)$, propositional systems were introduced. It should be noted that there are paraconsistent logics and that they are not paracomplete and vice versa; they are mathematically independent concepts. Today, many non-alethic systems are known, including the important annotated systems. We apply the Curry system concept to obtain algebraization of systems N_n^* $(1 \leq n < w)$.

8.2 The N_n* (1 ≤ n < ω) Non-Alethic Logic

In this section, we summarize a family of non-alethic predicate calculus N_n^* $(1 \leq n < w)$. For each n, $1 \leq n < w$, we have a different calculus symbolized by N_n^*. N_n^* $(1 \leq n < w)$ non-alethic calculus is a hybrid calculus of C_n^* $(1 \leq n < w)$ and P_n^* $(1 \leq n < w)$ calculus. Thus, such calculus has the following conditions:

(1) The Principle of Non-contradiction, in the form $\neg(A \wedge \neg A)$ is not generally valid;
(2) We cannot deduce any formula B from two contradictory propositions, A and $\neg A$.
(3) There are formulas A and $\neg A$ that are both false.
(4) The calculus must contain the classical predicate calculus's essential schemes and inference rules compatible with conditions (1) and (2) above.

J. M. Abe, *An Introduction to Curry Systems*, Intelligent Systems Reference Library 280, https://doi.org/10.1007/978-3-031-97810-4_8

The language of the calculus N_n^* ($1 \leq n < w$) is the same for all of them; let us denote by L. The language L is a first-order language whose primitive symbols are the following:

(1) Individual variables: an infinite denumerable set of symbols.
(2) logical connectives: $\neg$ (negation), $\wedge$ (conjunction), $\vee$ (disjunction), and $\rightarrow$ (implication).
(3) for each n, zero or more n-ary function symbols (n is a natural number).
(4) for each n, zero or more n-ary predicate symbols.
(5) Quantifiers: $\forall$ (universal) and $\exists$ (existential)
(6) auxiliary symbols: parentheses and commas.

For each n, the number of function symbols can be zero or not, finite or infinite. A 0-ary function symbol is called a constant. Also, the number of predicate symbols can be finite or infinite.

Formulas are defined in the usual way. In L we put:

Definition 2.1 Let A be a formula. A° abbreviates $\neg(A \wedge \neg A)$. A^{i} abbreviates $A^{\circ \cdots \circ}$, where the symbol $^{\circ}$ appears i times, $i \geq 1$. Thus, A^1 *is* A^o. We also write $A^{(i)}$ for $(A \wedge A^1 \wedge A^2 \wedge \ldots \wedge A^i)$.

Definition 2.2 Let A be a formula. We have that $A^\#$ abbreviates $A \wedge \neg A$. We also write for $A^{[\mathrm{i}]}$ for $A^\# \wedge A^{\#\#} \wedge A^{\#\#\#} \wedge \ldots \wedge A^{\#n}$, where $^{\#\mathrm{n}}$ means that the symbol $^\#$ appears n times.

Definition 2.3 We write $(A \leftrightarrow B)$ for $(A \rightarrow B) \wedge (B \rightarrow A)$ and $\neg^* A$ for $A^{(n)} \wedge \neg A$.

The postulates (axiom schemes and inference rule) of N_n^* ($1 \leq n < w$) are as follows: A, B, C denote any formulas.

($\rightarrow_1$) $A \rightarrow (B \rightarrow A)$
($\rightarrow_2$) $(A \rightarrow (B \rightarrow C)) \rightarrow ((A \rightarrow B) \rightarrow (A \rightarrow C))$
($\rightarrow_3$) $\frac{A, A \rightarrow B}{B}$ (MP (Modus Ponens))
($\wedge_1$) $A \wedge B \rightarrow A$
($\wedge_2$) $A \wedge B \rightarrow B$
($\wedge_3$) $A \rightarrow (B \rightarrow (A \wedge B))$
($\vee_1$) $A \rightarrow A \vee B$
($\vee_2$) $B \rightarrow A \vee B$
($\wedge_3$) $(A \rightarrow C) \rightarrow ((B \rightarrow C) \rightarrow ((A \vee B) \rightarrow C))$
($\neg_{10}$) $A^{[n]} \wedge B^{(n)} \rightarrow ((A \rightarrow B) \rightarrow ((A \rightarrow \neg B) \rightarrow \neg A)$
($\neg_4$) $(A^{(n)} \wedge B^{(n)}) \rightarrow ((A \wedge B)^{(n)} \wedge (A \wedge B)^{(n)} \wedge (A \rightarrow B)^{(n)})$
($\neg_9$) $(A^{[n]} \wedge B^{[n]}) \rightarrow ((A \wedge B)^{[n]} \wedge (A \wedge B)^{[n]} \wedge (A \rightarrow B)^{[n]})$
($\neg_{11}$) $A^{(n)} \rightarrow (A \rightarrow \neg\neg A) \wedge (A \rightarrow (\neg A \rightarrow B))$
($\neg_{12}$) $A^{[n]} \rightarrow (\neg\neg A \rightarrow A)$
(N) $A^{(n)} \vee A^{[n]}$
($\exists_1$) $A(t) \rightarrow \exists x A(x)$
($\exists_2$) $\frac{A(x) \rightarrow B}{\exists x A(x) \rightarrow B}$

($\forall_1$) $\forall x A(x) \to A(t)$

($\forall_2$) $\frac{B \to A(x)}{B \to \forall x A(x)}$

($\forall_3$) $\forall x(A(x))^{(n)} \to (\forall x A(x))^{(n)}$

($\exists_3$) $\forall x(A(x))^{(n)} \to (\exists x A(x))^{(n)}$

($\forall_4$) $\forall x(A(x))^{[n]} \to (\forall x A(x))^{[n]}$

($\exists_4$) $\forall x(A(x))^{[n]} \to (\exists x A(x))^{[n]}$

with the usual restrictions.

(C) If A and B are congruent formulas [2] or if one results from another by suppressing vacuous quantifiers, then $A \leftrightarrow B$ constitutes an axiom.

The N_w^* calculus is obtained from N^*_n by suppressing the axiom schemes ($\neg_{10}$), ($\neg_4$), ($\neg_9$), ($\neg_{11}$), ($\neg_{12}$), (N), ($\exists_3$), ($\exists_4$), ($\forall_4$) *and* ($\forall_4$).

Theorem 2.4 In N_n^* ($1 \leq n < w$), all valid schemes and rules of classical positive propositional logic are valid. In particular, the deduction theorem is valid N_n^* ($1 \leq n < w$). N_n^* ($1 \leq n < w$) contains the positive intuitionist logic.

Proof An immediate consequence of the postulates. Since Peirce's Law is valid in N_n^* ($1 \leq n < w$), they contain the positive intuitionist logic.

Theorem 2.5 In N_n^* ($1 \leq n < w$), we have:

Let A be a formula : $\vdash \neg A^{(n)} \to (\neg A)^{(n)}$

As the N_n^* ($1 \leq n < w$) calculus was built from the C_n^* ($1 \leq n < w$) calculus and the P_n^* ($1 \leq n < w$) calculus, the meanings of the expressions $A^{(n)}$ and $A^{[n]}$ are the same as previously exposed. It should be noted, then, that the calculus N_n^* ($1 \leq n < w$) has all the characteristics of C_n* and P_n*, being paraconsistent and paracomplete logics simultaneously. Furthermore, in these calculi, the connectives $\to$, $\wedge$, $\vee$ *and* $\neg^*$ have all the properties of the classic implication, conjunction, disjunction and negation, respectively, which together with the universal and existential quantifiers make the classical predicate calculus contained in N_n^*.

Theorem 2.6 In N_n^* ($1 \leq n < w$), the following schemes are not valid:

Let A be a formula : $A^o \wedge A^*, A \to \neg A^*, \neg A \ \wedge A^o, A^* \vee \neg A^*$

Theorem 2.7 In N_n^* ($1 \leq n < w$), the following schemes are not valid:

1. $A \wedge \neg A$
2. $\neg(A \vee B) \leftrightarrow (\neg A \wedge \neg B)$
3. $\neg(A \wedge B) \leftrightarrow (\neg A \vee \neg B)$
4. $\neg\neg A \to A$
5. $(A \wedge \neg A) \leftrightarrow (B \wedge \neg B)$

If N_0^* denotes the classical predicate calculus, then the hierarchy $N_0^*, N_1^*, N_2^*, \ldots, N_n^*, \ldots, N_w^*$ is such that N_{i+1}^* is strictly weaker than N_i^* for all ($1 \leq i < w$). N_w^* is the weakest calculus in the hierarchy. It is worth noting that we can extend N_i* to higher-order logics and strong theories that can be formalized in these logics.

Also, these logics have a double-valued semantics, which means they are sound and complete. Let us consider only the N_1^* calculus in order to fix ideas. N_1^* is a paraconsistent and paracomplete calculus, that is, non-alethic; thus, we can use it to manipulate a set of inconsistent and paracomplete formulas without immediate trivialization (by this, we mean that not all formulas are theorems).

It is easy to show this property to N_1^*; there are inconsistent (paracomplete) theories with models, and consequently, they are not trivial. In other words, N_1^* is the underlying logic of paraconsistent and paracomplete theories, that is, of non-alethic theories.

However, note that when we are only working with formulas that satisfy the Principle of Non-contradiction, then N_1^* reduces to C_1^*, and if we are considering formulas that only satisfy the Principle of Excluded Middle, N_1^* reduces to P_1*.

If we are working with both the Principle of Non-contradiction and the Principle of Excluded Middle, we obtain the classic calculus N_0^*.

8.3 N_1*-Monadic Algebras

In this session, we will study N_1^* monadic algebras; see [3]. Such algebraic systems constitute one of the possible algebraic versions of the monadic calculus N_1^* seen earlier.

Definition 3.1 Let A be an N_1-algebra. Let $\exists$ and $\forall$ be operators over A. The pair $(\exists, \forall)$ is called a quantifier over A if:

1. $\exists 0 \equiv 0$;
2. $x \leq \exists x$;
3. $\exists(x \wedge y) \equiv \exists x \wedge \exists y$;
4. $\exists\exists x \equiv \exists x$;
5. $\exists(\exists x)^* \equiv (\exists x)^*$;
6. $\exists(x \wedge \exists y) \equiv \exists x \wedge \exists y$;
7. $\forall 1 \equiv 1$;
8. $\forall x \leq x$;
9. $\forall(x \wedge y) \equiv \forall x \wedge \forall y$;
10. $\forall\forall x \equiv \forall x$;
11. $\forall(\forall x)^* \equiv (\forall x)^*$

We suppose in the above definition that, if $x \equiv y$, then $\exists x \equiv \exists y$ and $\forall x \equiv \forall y$. $\exists$ is called existential quantifier on A and $\forall$ is called universal quantifier on A. The pair $[A, (\exists, \forall)]$ is called a monadic Curry algebra N_1^* or a N_1^* -monadic algebra (or N_1^*-algebra).

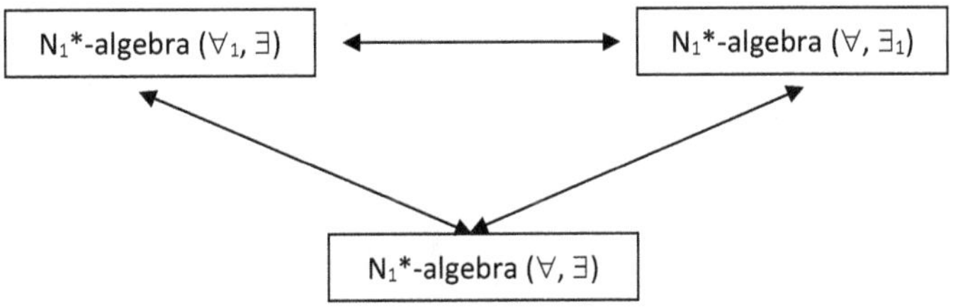

Fig. 8.1 Obtaining N_1*-monadic algebras

Given a Curry algebra P_1, let's assume that there is a universal quantifier defined on it, i.e., a structure $[A, \forall]$ such that conditions (7)—(11) above are satisfied. If we define $\exists_1 x =_{Def.} (\forall x^*)^*$, then $\exists_1$ is an existential quantifier (i.e., satisfying (1)—(6) and the structure composed by $[A, (\exists_1, \forall)]$ is a N_1^*-monadic algebra. Also, we can get a monadic algebra considering an existential quantifier $\exists$ on a Curry algebra P_1 satisfying conditions (1)—(6) of the above definition and defining a universal quantifier (i.e., satisfying (7)—(11) as $\forall_1 x =_{Def.} (\exists x^*)^*$. Then the structure composed by $[A, (\exists, \forall_1)]$ is a N_1^* -monadic algebra. Given a Curry algebra N_1, in general, the algebras obtained $[A, (\exists_1, \forall)]$ and $[A, (\exists, \forall_1)]$ are not isomorphic. Also, given a N_1^*-monadic algebra $[A, (\exists, \forall)]$, define the new quantifiers $\exists_1$ and $\forall_1$. In general, we have $\exists_1 \neq \exists$. and $\forall_1 \neq \forall$ (Fig. 8.1).

Let N be a N_1-algebra and $A = \{(x_1, x_2, \ldots, x_n) | x_i \in N, i = 1, 2, \ldots, n\}$. Let us suppose that if $x, y \in A$, then $x \wedge y \in A$ and $x' \in A$, and if $(x_1, x_2, \ldots, x_n), (y_1, y_2, \ldots, y_n) \in A$ we define $(x_1, x_2, \ldots, x_n) \equiv (y_1, y_2, \ldots, y_n)$ iff $x_i \equiv y_i, i = 1, 2, \ldots, n$. Also, we put $\overline{x} = \vee_{i=1}^{n} x_i, \underline{x} = \wedge_{i-1}^{n} x_i$, and we assume that $\underbrace{(\overline{x}, \ldots, \overline{x})}_{n-times}, \underbrace{(\underline{x}, \ldots, \underline{x})}_{n-times} \in A$. If we define $\exists x = (\overline{x}, \ldots, \overline{x})$ and $\forall x = (\underline{x}, \ldots, \underline{x})$, then A is a N_1^*-monadic algebra.

A more useful example of N_1*-monadic algebra is the following. Let N be a N_1-algebra, a set $K^1 \neq \emptyset$, and N^K is the set of all functions of K into N. Let A be the set such that: (i) A is a N_1-algebra with respect the pointwise operations, and (ii) if $x \in A$, then the range of x has a supremum $\overline{x}$ and a infimum $\underline{x}$ in N, and the functions that take the value $\overline{x}$ at each point of K and $\underline{x}$ at each point of K are in A. If $\exists x$ and $\forall x$ are defined to be those functions, then A becomes a N_1^*- monadic algebra. Every N_1^*- monadic algebra obtained in this way is called an N-valued functional algebra with domain K.

8.4 The N_1*-Monadic Algebras and Monadic Algebras

Now we will discuss some of the results of the algebraic structures that originated in our discussion. The matter is very rich; however, we will only concern ourselves with some.

Theorem 4.1 In a N_1^*-monadic algebra $[A, (\exists, \forall)]$, the structure composed by the underlying set and by operations $\wedge$, $\vee$, *, $\exists$, *and* $\forall$ is a (pre) monadic algebra. If we pass to the quotient by the fundamental relation $\equiv$, we obtain a monadic algebra in the usual sense [4].

Definition 4.2 Let $[A, (\exists, \forall)]$ be a N_1^* -monadic algebra, and $[A, \equiv, \leq, \rightarrow, {}^*, \exists, \forall]$ the monadic algebra obtained as in the above theorem. Any monadic algebra that is isomorphic to the quotient algebra of $[A, \equiv, \leq, \rightarrow, {}^*, \exists, \forall]$ by $\equiv$ is called a monadic algebra *associated with the* N_1^*-monadic algebra.

Hence, we can establish the following representation theorems for N_1^*-monadic algebras.

Theorem 4.3 If N is a N_1^*-monadic algebra, then for its associated monadic algebra A, there exists a set X, and there exists a Boolean an algebra B, such that (i) A is isomorphic to a B-valued functional algebra A' with domain X, and (ii) for every element p of A' there exists a point x in X with $p(x) = \exists p(x)$.

Theorems 4.2 and 4.3 show us that N_1^*-monadic algebras constitute an interesting generalization of monadic algebras. Here, there is an open problem. How many non-isomorphic monadic algebras associated with an N_1^* -monadic algebra are there?

References

1. da Costa, Newton C.A.: Logics that are both paraconsistent and paracomplete. Atti della Accademia Nazionale dei Lincei. Classe di Scienze Fisiche, Matematiche e Naturali. Rendiconti Lincei. Matematica e Applicazioni 83.1 (1989): 29–32. <http://eudml.org/doc/287311>
2. Kleene, S.C.: Introduction to metamathematics, Ishi press, Japan, (2009). ISBN-10: 0923891579, ISBN-13: 978-0923891572, p. 572
3. Abe, J.M., Nakamatsu, K., Akama, S.: Monadic curry system N_1*, lecture notes in computer science, ISSN 0302–9743, Springer-Verlag, Berlin/Heidelberg, Alemanha. LNAI **6278**, 143–153 (2010). https://doi.org/10.1007/978-3-642-15393-8_17
4. Halmos, P.H.: Algebraic logic, dover publications (2018). ISBN-10: 0486828042, ISBN-13: 978–0486828046
5. Abe, J.M., Nakamatsu, K., Akama, S.: Monadic curry system N1*, Lecture Notes in Computer Science, ISSN 0302-9743, Springer-Verlag, Berlin/Heidelberg, Germany. LNAI **6278**, 143–153 (2010). https://doi.org/10.1007/978-3-642-15393-8_17

Chapter 9
Pτ-Algebras and Monadic Curry Algebras Qτ

9.1 Introduction

The annotated logic first appeared as programming logic, and the underlying logic, known as annotated paraconsistent logic, was subsequently studied; see Abe et al [1–3].

The annotated logics are two-sorted logics, and in their syntax, a lattice is associated with them, which plays a crucial role in the structure of the logic. The annotated logics are paraconsistent and, in general, paracomplete and non-alethic logics according to the lattice associated. The annotated logics are surprisingly functional logics. From the application point of view, it can be said that the annotated logics is the "most certain" paraconsistent logic that has been discovered so far. A detailed account of annotated logics is to be found in [1, 3].

One question arises naturally: what algebraic version of these logics? In this chapter, we introduce the Curry Algebras Pτ that algebraises the annotated systems Pι and an algebraic version of monadic predicate annotated logics Qτ, when the lattice τ is finite. Some basic results are presented. Curry systems constitute an up-and-coming field of research, but few people have given attention to this theme.

9.2 The Annotated Logics Pτ

In this section, we summarize the annotated logics Pτ. References for this paragraph are [1, 3]. Pτ is a family of paraconsistent propositional logics. They are defined as follows: throughout this work, $\tau = < |\tau|, \leq >$ will be some arbitrary, but fixed finite lattice of truth-values (many times we identify τ with its underlying set |τ|). The least element of τ is denoted by ⊥, while its greatest element by ⊤. We also assume that there is a fixed unary operator ~: |τ| → |τ| which constitutes the "meaning" of our

J. M. Abe, *An Introduction to Curry Systems*, Intelligent Systems Reference Library 280, https://doi.org/10.1007/978-3-031-97810-4_9

negation. $\vee$ and $\wedge$ denote, respectively, the least upper bound and the greatest lower bound operators of τ.

The language of Pτ is composed of the following primitive symbols.

1. Propositional symbols p, q, ... (with or without numerical subscripts)
2. Each member of τ is an annotation constant: μ, λ, …
3. The connectives $\neg$, $\wedge$, $\vee$, and $\rightarrow$
4. Auxiliary symbols: parentheses.

Formulas are defined as follows:

1. If p is a propositional symbol and λ is an annotation constant, then p_λ is an (atomic) formula.
2. If A and B are formulas, then $(\neg A)$, $(A \wedge B)$, $(A \vee B)$, and $(A \rightarrow B)$ are formulas.

Among several intuitive readings, the atomic formula p_λ can be read as *it is believed that p_λ's truth-value is at least λ.*

We introduce some definitions.

1. The symbol of equivalence ' $\leftrightarrow$ ' is introduced as usual.
2. Let A be a formula. $\daleth A = A \rightarrow ((A \rightarrow A) \wedge \neg(A \rightarrow A))$ is called the *strong negation of A.*
3. $\underbrace{\neg \ldots \neg}_{k-times} p_\lambda = \neg^k p_\lambda$ is called a *hyper-literal*. A formula other than hyper-literals is called a *complex formula.*

Definition 2.1 An *interpretation I* is a function I: $P \rightarrow |\tau|$. To each interpretation I, we associate a valuation V_I: $F \rightarrow \mathbf{2}$, where P is the set of propositional symbols, F is the set of all formulas, and $\mathbf{2} = \{0, 1\}$, defined as follows:

1. If p_λ is an atomic formula, then:
   (a) $V_I(p_\lambda) = 1$ iff $I(A) \geq \lambda$
   (b) $V_I(p_\lambda) = 0$ otherwise.
   (c) $V_I(\neg^k p_\lambda) = V_I(\neg^{k-1} p_{\sim\lambda})$, $k \geq 1$
2. If A and B are formulas, then:
   (a) $V_I(A \wedge B) = 1$ iff $V_I(A) = V_I(B) = 1$
   (b) $V_I(A \vee B) = 1$ iff $V_I(A) = 1$ or $V_I(B) = 1$
   (c) $V_I(A \rightarrow B) = 1$ iff $V_I(A) = 0$ or $V_I(B) = 1$
3. If A is a complex formula, then
   $V_I(\neg A) = 1 - V_I(A)$.

Now, we present an axiomatization of Pτ. In the sequel, A, B, and C are formulas whatsoever, F and G are complex formulas, p is a propositional symbol, and μ, μ_j, $1 \leq j \leq n$, are annotational constants.

$(\rightarrow_1)$ $A \rightarrow (B \rightarrow A))$
$(\rightarrow_2)$ $(A \rightarrow (B \rightarrow C)) \rightarrow ((A \rightarrow B) \rightarrow (A \rightarrow C))$

($\rightarrow_3$) $((A \rightarrow B) \rightarrow A) \rightarrow A$
($\rightarrow_4$) $A, A \rightarrow B/B$ (Modus Ponens, abbreviated by MP)
($\wedge_1$) $(A \wedge B) \rightarrow A$
($\wedge_2$) $(A \wedge B) \rightarrow B$
($\wedge_3$) $A \rightarrow (B \rightarrow (A \wedge B))$
($\vee_1$) $A \rightarrow (A \vee B)$
($\vee_2$) $B \rightarrow (A \vee B)$
($\vee_3$) $(A \rightarrow C) \rightarrow ((B \rightarrow C) \rightarrow ((A \vee B) \rightarrow C))$
($\neg_1$) $(F \rightarrow G) \rightarrow ((F \rightarrow \neg G) \rightarrow \neg F)$
($\neg_2$) $F \rightarrow (\neg F \rightarrow A)$
($\neg_3$) $F \vee \neg F$
(τ_1) $p_\perp$
(τ_2) $(\neg^k p_\mu) \leftrightarrow (\neg^{k-1} p_{\sim\mu}), k \geq 1$
(τ_3) $p_\mu \rightarrow p_\lambda$, where $\mu \geq \lambda$
(τ_4) $p_{\mu 1} \wedge p_{\mu 2} \wedge \ldots p_{\mu n} \rightarrow p_\lambda$, where $\lambda = \bigvee_{i=1}^{n} \mu_i$

We can introduce usual syntactical concepts as proof, deduction from hypotheses, etc. Abe [1] proved that the axiomatization above is sound and complete concerning the semantics discussed (when the associated lattice is finite).

Some remarkable results of the systems Pτ are:

(a) The set of formulas together with the connectives $\urcorner\!\!\urcorner$, $\wedge$, $\vee$, and $\rightarrow$ has all properties of the Classical Propositional logic (see Theorem 4 below).
(b) The set of all complex formulas with the connectives $\neg$, $\wedge$, $\vee$, and $\rightarrow$ also has all properties of the classical propositional logic. So, the inconsistencies appear among hyper-literals, or equivalently, among atomic formulas. This nice property, so in applications, for instance, in databases, the conflicting information is among hyper-literals; we hardly need inconsistencies among complex formulas.
(c) Pτ is a paraconsistent logic and, in general, paracomplete and non-alethic logics.

Definition 2.2 In Pτ: we define $A \leq B$ by setting $\vdash A \rightarrow B$ and $A \equiv B$ by setting $A \leq B$ and $B \leq A$.

Theorem 2.3 $\leq$ is a quasi-order and $\equiv$ is an equivalence relation.

Theorem 2.4 In general, in Pτ, $\neg$ is not compatible with the equivalence relation $\equiv$, and we have $\vdash (A \wedge \urcorner\!\!\urcorner A) \rightarrow B$.

Theorem 2.5 The strong negation possesses all properties of the classical negation; for instance, let A, B, and C be formulas.

$\vdash (A \rightarrow B) \rightarrow ((A \rightarrow \urcorner\!\!\urcorner B) \rightarrow \urcorner\!\!\urcorner A)$
$\vdash A \rightarrow (\urcorner\!\!\urcorner A \rightarrow B)$.
$\vdash (A \wedge \urcorner\!\!\urcorner A) \rightarrow B.A$
$\vdash A \rightarrow \urcorner\!\!\urcorner\urcorner\!\!\urcorner A$.
$\vdash \urcorner\!\!\urcorner\urcorner\!\!\urcorner A \rightarrow A$.

9.3 Curry Algebras Pτ

The algebraic structures considered here follow previous discussions. Let us consider a non-empty set S and a finite lattice (in the usual sense) $\tau = (|\tau|, \leq)$, together with an operator~: $|\tau| \to |\tau|$. Let S^* be a set of all pairs (p, λ) where $p \in S$ and $\lambda \in |\tau|$. Consider now the set $S^* \cup \{\neg, \wedge, \vee, \to\}$. Let S^{**} be the smallest algebraic structure freely generated by the set $S^* \cup \{\neg, \wedge, \vee, \to\}$ (we do not write the details; it follows the usual method to obtain free structures). Elements of S^{**} are classified into two categories: *hyper-literal elements* are of the form $\neg^k(p, \lambda)$, and *complex elements* are the remaining elements of S^{**}.

Now we introduce the concept of a Curry algebra Pτ.

Definition 3.1 A Curry algebra Pτ (abbreviated by Pτ-algebra) is a structure

$R\tau = [R, (|\tau|, \leq, \sim), \equiv, \leq, \to, \neg]$ and, for $p \in R$, $a \in R^{**}$, and x, y complex elements of R^{**}:

1. R^{**} is a classical implicative lattice with a greatest element **1**.
2. $\neg$ is a unary operator $\neg: R^{**} \to R^{**}$.
3. $x \to y \leq (x \to \neg y) \to \neg x$
4. $x \leq \neg x \to a$
5. $(p, \bot) \equiv \mathbf{1}$ ($\bot$ is the bottom element of τ)
6. $x \vee \neg x \equiv \mathbf{1}$
7. $\neg^k(p, \lambda) \equiv \neg^{k-1}(p, \sim\lambda), k \geq 1$
8. $\mu \leq \lambda$ then $(p, \mu) \leq (p, \lambda)$
9. $(p, \lambda_1) \wedge (p, \lambda_2) \wedge \ldots \wedge (p, \lambda_n) \leq (p, \lambda)$, where $\lambda = \vee_{i=1}^{n} \lambda_i$

Theorem 3.2 A Pτ-algebra is distributive and has a greatest and first element.

Definition 3.3 Let x be an element of a Pτ-algebra. We put.

$$x^* = x \to ((x \to x) \wedge \neg(x \to x)).$$

Theorem 3.4 In a Pτ-algebra, x^* is a Boolean complement of x, so $x \vee x^* \equiv 1$ and $x \wedge x^* \equiv 0$.

Theorem 3.5 In a Pτ-algebra, the structure composed by the underlying set and by operations $\wedge$, $\vee$, and $*$ is a (pre) Boolean algebra. If we pass to the quotient by the fundamental relation $\equiv$, we obtain a Boolean algebra in the usual sense.

Definition 3.6 Let $[R, (|\tau|, \leq, \sim), \equiv, \leq, \to, \neg]$ be a Pτ-algebra, and $[R, (|\tau|, \leq, \sim), \equiv, \leq, \to, {}^*]$ the Boolean algebra obtained as in the above theorem. Any Boolean algebra that is isomorphic to the quotient algebra of $[R, (|\tau|, \leq, \sim), \equiv, \leq, \to, {}^*]$ by $\equiv$ is called Boolean algebra *associated with the* Pτ-algebra.

Hence, we can establish the representation theorems for Pτ-algebras.

Theorem 3.7 (Representation Theorem) Any Pτ-algebra is associated with a field of sets. Moreover, any Pτ-algebra is associated with the field of sets simultaneously open and closed of a totally disconnected compact Hausdorff space.

Proof. It follows the above definition and Stone's classical representation theorems for Boolean algebras.

Other methods exist for extracting Boolean algebras from Pτ-algebras. There is another natural Boolean algebra associated with a Pτ-algebra.

Definition 3.8 Let $[R, (|\tau|, \leq, \sim), \equiv, \leq, \rightarrow, \neg]$ be a Pτ-algebra. By *RC* we indicate the set of all complex elements of $[R, (|\tau|, \leq, \sim), \equiv, \leq, \rightarrow, \neg]$. Then the structure $[RC, (|\tau|, \leq, \sim), \equiv, \leq, \rightarrow, \neg]$ constitutes a (pre) Boolean algebra, which we call as *Boolean algebra c-associated with* the Pτ-algebra $[R, (|\tau|, \leq, \sim), \equiv, \leq, \rightarrow, \neg]$.

Thus, we obtain a second Representation theorem for Pτ-algebras.

Theorem 3.9 Any Pτ-algebra is *c-associated* with a field of sets. Moreover, any Pτ-algebra is *c-associated* with the field of sets simultaneously open and closed of a totally disconnected compact Hausdorff space.

Theorems 3.7 and 3.9 show us that Pτ-algebras constitute interesting generalizations of Boolean algebras. We finish this work with some open problems. How many non-isomorphic Boolean algebras associated with a Pτ-algebra are there? Furthermore, how many non-isomorphic Boolean algebras are c-associated with a Pτ-algebra? Establish connections between associated and c-associated algebras; see Abe [6].

9.4 Monadic Curry Algebras Qτ

In this session, we introduce the concept of monadic Curry algebras Qτ that algebraises the annotated systems Qτ extensively studied by Abe [1]. Some basic initial results are presented.

Now, we shall describe an axiomatic system called *A*τ whose underlying language is *L*: *A*, *B*, *and C* are any formulas whatsoever, *F*, *G* are complex formulas, and $p_\lambda t_1 \ldots t_n$ an annotated atom. *A*τ consists of the following postulates (axiom schemes and primitive rules of inference) with the usual restrictions:

The postulates of Qτ are those of postulates of Pτ plus the following postulates:

($\exists_1$) $A(t) \rightarrow \exists x A(x)$

($\exists_2$) $\frac{A(x)\rightarrow B}{\exists x A(x)\rightarrow B}$

($\forall_1$) $\forall x A(x) \rightarrow A(t)$

($\forall_2$) $\frac{B\rightarrow A(x)}{B\rightarrow \forall x A(x)}$

(τ_1) $p_\perp t_1 \ldots t_n$

(τ_2) $\neg^k p_\lambda t_1 \ldots t_n \rightarrow \neg^{k-1} p_{\sim\lambda} t_1 \ldots t_n, k \geq 1$

(τ_3) $p_\lambda t_1 \ldots t_n \rightarrow p_\mu t_1 \ldots t_n, \lambda \geq \mu$

(τ_4) $p_{\lambda 1} t_1 \ldots t_n \wedge p_{\lambda 2} t_1 \ldots t_n \wedge \ldots \wedge p_{\lambda m} t_1 \ldots t_n \rightarrow p_\lambda t_1 \ldots t_n$, where $\lambda = \vee_{i=1}^m \lambda_i$

$(=_1)$ $x = x$
$(=_2)$ $x = y \to \big(A[x] \leftrightarrow A\big[y\big]\big)$

Definition 4.1 Let A be a Pτ-algebra. Let $\exists$ and $\forall$ be operators on A. $(\exists, \forall)$ is called a quantifier on A if:

1. $\exists 0 \equiv 0$
2. $x \le \exists x$
3. $\exists(x \vee y) \equiv \exists x \vee \exists y$
4. $\exists\exists x \equiv \exists x$
5. $\exists(\exists x)^* \equiv (\exists x)^*$
6. $\exists(x \wedge \exists y) \equiv \exists x \wedge \exists y$
7. $\forall 1 \equiv 1$
8. $\forall x \le x$
9. $\forall(x \vee y) \equiv \forall x \vee \forall y$
10. $\forall\forall x \equiv \forall x$
11. $\forall(\forall x)^* \equiv (\forall x)^*$

We suppose in the above definition that, if $x \equiv y$, then $\exists x \equiv \exists y$ and $\forall x \equiv \forall y$. $\exists$ is called existential quantifier on A and $\forall$ is called universal quantifier on A. The pair $< A, (\exists, \forall) >$ is called a monadic Curry algebra Qτ or a Qτ-monadic algebra (or Qτ-algebra).

Let C be a Pτ-algebra and $A = \{(x_1, x_2 \ldots, x_n) \mid x_i \in C,\ i = 1, 2, \ldots, n\}$. Let us suppose that if $x, y \in A$, then $x \vee y \in A$ and $\neg x \in A$, and if $(x_1, x_2, \ldots, x_n), (y_1, y_2, \ldots, y_n) \in A$ we define $(x_1, x_2, \ldots, x_n) \equiv (y_1, y_2, \ldots, y_n)$ iff $x_i \equiv y_i$, $i = 1, 2, \ldots, n$. Also, we put $\overline{x} = \bigvee_{i=1}^{n} x_i$, $\underline{x} = \bigwedge_{i=1}^{n} x_i$, and assume that $\underbrace{(\overline{x}, \ldots \overline{x})}_{n-times}, \underbrace{(\underline{x} \cdots, \underline{x})}_{n-times} \in A$. If we define $\exists x = (\overline{x}, \ldots, \overline{x})$ and $\forall x = (\underline{x}, \cdots \underline{x})$, then A is a Qτ-monadic algebra.

A more useful example of Qτ-monadic algebras is the following. Let C be a Pτ-algebra, a set $K \neq \emptyset$, and C^K is the set of all functions of K into C. Let A be the set such that: (i) A is a Pτ-algebra concerning the pointwise operations, and (ii) if $x \in A$, then the range of x has a supremum $\overline{x}$ and an infimum $\underline{x}$ in C, and the functions that take the value $\overline{x}$ at each point of K and $\underline{x}$ at each point of K are in A. If $\exists x$ and $\forall x$ are defined as those functions, then A becomes a Qτ-monadic algebra. Every Qτ-monadic algebra obtained in this way is called a C-valued functional algebra with domain K.

9.5 Qτ-Monadic Algebras and Monadic Algebras

Now we discuss some aspects of the algebraic structures that originated in our discussion. The matter is very rich, and we will only be concerned about some of them. As in paragraph 4, given a Pτ-algebra, it generates (at least) two ways to obtain (classical)

Boolean algebras: associated to and c-associated to. These two algebras generate two corresponding (classical, in the sense of [7]) monadic algebras. Each of these extensions, namely Qτ-monadic associated Pτ-algebra and Qτ-monadic c-associated Pτ-algebra, has a representation theorem in the following form.

In the sequel, we use the definitions and conventions in paragraph 4.

Theorem 5.1 In a Qτ-algebra, the structure composed by the underlying set and by operations $\wedge$, $\vee$, $*$, $\exists$, and $\forall$ is a (pre) monadic algebra. If we pass to the quotient by the fundamental relation $\equiv$, we obtain a monadic algebra in the usual sense [8].

Definition 5.2 Let $(R, (|\tau|, \leq, \sim), \equiv, \leq, \rightarrow, \neg, \exists, \forall)$ be a Qτ-algebra, and $(R, (|\tau|, \leq, \sim), \equiv, \leq, \rightarrow, {}^*, \exists, \forall)$ the monadic algebra obtained as in the above theorem. Any monadic algebra that is isomorphic to the quotient algebra of $(R, (|\tau|, \leq, \sim), \equiv, \leq, \rightarrow, {}^*, \exists, \forall)$ by $\equiv$ is called a monadic algebra *associated with the* Qτ-algebra.

Hence, we can establish the following representation theorems for Qτ-algebras.

Theorem 5.3 If C is a Qτ-algebra, then for its associated monadic algebra A, there exists a set X, and there exists a Boolean algebra B, such that (i) A is isomorphic to a B-valued functional algebra A' with domain X, and (ii) for every element p of A' there exists a point x in X with $p(x) = \exists p(x)$.

A second way of extracting monadic algebras from Qτ-algebras exists. There is another natural monadic algebra associated with a Qτ-algebra.

Definition 5.4 Let $(R, (|\tau|, \leq, \sim), \equiv, \leq, \rightarrow, \neg, \exists, \forall)$ be a Qτ-algebra. By RC we indicate the set of all complex elements of $(R, (|\tau|, \leq, \sim), \equiv, \leq, \rightarrow, \neg, \exists, \forall)$ Then the structure $(RC, (|\tau|, \leq, \sim), \equiv, \leq, \rightarrow, \neg, \exists, \forall)$ constitutes a (pre) monadic algebra, which we call as *monadic algebra c-associated with* the Qτ-algebra $(R, (|\tau|, \leq, \sim), \equiv, \leq, \rightarrow, \neg, \exists, \forall)$.

Thus, we obtain a second representation theorem for Qτ-algebras.

Theorem 5.5 If C is a Qτ-algebra, then for its c-associated monadic algebra A, there exists a set X, and there exists a Boolean algebra B, such that (i) A is isomorphic to a B-valued functional algebra A' with domain X, and (ii) for every element p of A' there exists a point x in X with $p(x) = \exists p(x)$.

Theorems 5.3 and 5.5 show us that Qτ-algebras constitute interesting generalizations of monadic algebras. We finish this work with some open problems. How many non-isomorphic monadic algebras are associated with a Qτ-algebra? Moreover, how many non-isomorphic monadic algebras are c-associated with a Qτ-algebra? Establish connections between associated and c-associated monadic Qτ-algebras. All these questions are to be considered in forthcoming works.

References

1. Abe, J.M.: Fundamentos da lógica anotada (Foundations of annotated logics), in Portuguese, Doctor thesis, University of São Paulo, São Paulo, p. 135 (1992)
2. Abe, J.M., Akama, S.: Annotated logics Qτ and ultraproducts, Logique et Analyse, Nouvelle Série, vol. 40, No. 160, pp. 335–343. Peeters Publishers, (1997). https://doi.org/10.2307/44084610 https://www.jstor.org/stable/44084610
3. Abe, J.M., Akama, S., Nakamatsu, K.: Introduction to annotated logics-foundations for paracomplete and paraconsistent reasoning, series title intelligent systems reference library, vol. 88, 1st edn, p. 190 Publisher Springer International Publishing, Copyright Holder Springer International Publishing Switzerland, eBook ISBN 978-3-319-17912-4, https://doi.org/10.1007/978-3-319-17912-4, Hardcover ISBN 978-3-319-17911-7, Series ISSN 1868-4394, (2015)
4. Abe J.M., Akama S., Nakamatsu K.: Monadic curry algebras Qτ. In: Apolloni, B., Howlett, R.J., Jain, L. (eds) Knowledge-Based Intelligent Information and Engineering Systems. KES 2007. Lecture Notes in Computer Science, vol. 4693, Springer, Berlin, Heidelberg. Print ISBN978–3–540–74826–7, Online ISBN978–3–540–74827–4, https://doi.org/10.1007/978-3-540-74827-4_112, https://doi.org/10.1007/978-3-540-74827-4_112 (2007)
5. Goldblatt, R.: Axiomatising the logic of computer programming. Lecture notes in computer science, vol. 130, Springer-Verlag, Berlin, Heidelberg, and New York, pp. 304 (1982)
6. Abe, J.M.: Curry algebras Pτ, Logique et Analyse, Nouvelle Série, vol. 41, No. 161/163, pp. 5–15, Peeters Publishers. https://doi.org/10.2307/44084618, https://www.jstor.org/stable/44084618 (1998)
7. da Costa, N.C.A.: On the theory of inconsistent formal systems notre dame. J. of Formal Logic **15**, 497–510 (1974). https://doi.org/10.1305/ndjfl/1093891487
8. Henkin, L.: Le Structure Algèbrique des Théories Mathématiques, Gauthiers-Villars, 1st edn, Gauthier-Villars, vol. 8, pp. 52 (1956)
9. Abe, J.M.: An introduction to the theory of curry systems. Sci. Math. Jpn. **76**(2), 175–194. ISSN 1346-0862 (2013)

Chapter 10
Alternative Algebraization of Classic Logic

In this chapter, we show how the algebraization of classical propositional calculus, without passing to the quotient, using 'pre-structures,' is convenient for treating certain fundamental concepts of classical propositional logic. Thus, pre-structures seem convenient even in classical logic when passing to the quotient using appropriate and significant equivalences is possible. N.C.A. da Costa recognized this phenomenon many years ago, independently of M. Eytan. In what follows, we base our exposition on Eytan [1].

Eytan provides an algebraic interpretation of logical methods, particularly focusing on the tableau method introduced by Smullyan and the Hintikka sets used in model-theoretic semantics. The paper situates these concepts within classical propositional logic while attempting to highlight their algebraic structures. Eytan observes the traditional approaches in Boolean algebra for their limitations in offering an adequate algebraic depiction of these logical frameworks, asserting that a deeper exploration of these systems is necessary to capture the richness of logical deduction.

10.1 Boolean's Pre-Algebra

In §6, Chap. 1 it was observed that Boolean's concept of pre-algebra. In this paragraph, we develop the study of this structure.

Theorem 1.1 Let $[A, \equiv, \wedge, \vee, ', 0, 1]$ be a Boolean pre-algebra. Let us define the binary operation $\rightarrow$ over $A : x \rightarrow y =_{\text{Def}} x' \vee y$. Let also be $\leq$ the pre-order relation obtained in Theorem 7, Ch. I. The following properties then arise:

(A1) The pre-order $\leq$ is compatible with the operators $x \wedge \_$, $\_ \wedge x$, $x \vee \_$, $\_ \wedge x$, and $x \rightarrow -$.

(A2) For all $x \in A$, $x \leq 1$.

(A3) For all $x, y \in A$, we have $x \wedge y \leq x, y$

J. M. Abe, *An Introduction to Curry Systems*, Intelligent Systems Reference Library 280, https://doi.org/10.1007/978-3-031-97810-4_10

(A4) For all $x, y, z \in A$, if $z \leq x$ and $z \leq y$, then $z \leq x \wedge y$
(A5) For all $x, y \in A$, $(y \rightarrow x) \wedge y \leq x$.
(A6) For all $x, y, z \in A$, if $z \wedge y \leq x$, then $z \leq y \rightarrow x$
(A7) For all $x \in A$, $0 \leq x$
(A8) For all $x \in A$, $x \leq x \vee y$ and $y \leq x \vee y$
(A9) For all $x, y, z \in A$, if $x \leq z$ *and* $y \leq z$, then $x \vee y \leq z$
(A10) For all $x, y \in A$, $(x \rightarrow 0) \leq x'$ *and* $x' \leq (x \rightarrow 0)$
(A11) For all $x \in A$, $x'' \leq x$
(A12) For all $x, y \in A$, if $x \leq y$, then $y' \leq x'$
(A13) For all $x, y \in A$, $(x \vee y)' \leq x' \wedge y'$ *and* $x' \wedge y' \leq (x \vee y)'$
(A14) For all $x, y \in A$, $(x \wedge y)' \leq x' \vee y'$ *and* $x' \vee y' \leq (x \wedge y)'$
(A15) For all $x, y \in A$, $(x \rightarrow y) \leq x' \vee y$ *and* $x' \vee y \leq (x \rightarrow y)$

Conversely, if a Curry system $[A, \equiv, \rightarrow, \wedge, \vee, ', \leq, 0, 1]$ where:

(1) A is a non-empty set.
(2) 0 and 1 are two elements distinguished from A.
(3) $\rightarrow$, $\wedge$, $\vee$ are binary operations on A.
(4) $'$ is an operator in A.
(5) $\leq$ is a pre-order on A.

It satisfies the previous A1–A15 properties, so A is a Boolean pre-algebra.
Proof. Immediate.

Theorem 1.2 The classical propositional calculus can be provided with a Boolean pre-algebra structure.
Proof. It is immediate.

Definition 1.3 Let $[B, \equiv, \rightarrow, \wedge, \vee, ', \leq, 0, 1]$ and $[B^*, \equiv^*, \rightarrow^*, \wedge^*, \vee^*, '^*, \leq^* 0^*, 1^*]$ Boolean's pre-algebras. A homomorphism of B in B^* is a function f of B in B^* such that it preserves algebraic operations, pre-order, and distinguished elements, that is, such that for any $x, y \in B$, we have:

(H_1) $f(x \rightarrow y) \equiv^* f(x) \rightarrow^* f(y)$
(H_2) $f(x \wedge y) \equiv^* f(x) \wedge^* f(y)$
(H_3) $f(x \vee y) \equiv^* f(x) \vee^* f(y)$
(H_4) $f(x') \equiv^* f(x)'^*$
(H_5) if $x \leq y$, then, $f(x) \leq^* f(y)$
(H_6) $f(0) \equiv^* 0^*$
(H_7) $f(1) \equiv^* 1^*$

Definition 1.4 Let B be a Boolean pre-algebra. A non-empty subset F of support B is called a filter if:

(F_1) if $x, y \in F$, then $x \wedge y \in F$
(F_2) For all $x, y \in B$, if $x \in F$ and $x \leq y$, then $y \in F$.

Theorem 1.5 Let $[B, \equiv, \rightarrow, \wedge, \vee, \neg, 0, 1]$ be Boolean's pre-algebra. We then have:

(1) Every filter can be extended to an ultrafilter.
(2) A non-empty subset F of B is a filter if and only if:

   (a) $1 \in F$
   (b) if $x \in F$ and $x \to y \in F$, then $y \in F$

(3) If f: $B \to B^*$ is a homomorphism and F is a filter of B, then f (F) is a filter of B.
(4) For every filter F of B, the conditions are equivalent:
(a) F is an ultrafilter.
(b) F is a prime filter.
(c) F is a filter and for all $x \in B$, $x \in F$ or $\neg x \in F$, but not simultaneously.
(d) For all $x \in B$, precisely one of the elements x, $\neg x$ belongs to F.
(e) B/F $= 2$, where $2 = \{0, 1\}$ is Boolean's pre-algebra with two elements and $B/ \equiv$ the pre-algebra obtained by passing to the quotient B by the equivalence relation $x \equiv y \Leftrightarrow x \to y \in F$ and $y \to x \in F$.

Definition 1.6 The base of a filter F is called a non-empty subset Γ of A (support of A) such that:

(B) If $x, y \in \Gamma$, then there is $z \in \Gamma$ such that $z \leq x \wedge y$.

Theorem 1.7 A subset Γ of B is a base of a filter if, and only if the set of elements $x \in B$ such that y $\leq x$ for a $y \in \Gamma$, is a filter F. Such a filter is proper if $0 \notin \Gamma$.

Proof. ($\Rightarrow$). Suppose Γ is the basis of a filter. Let x, z be such that $y \leq x$ and $y' \leq z$, where y, $y' \in \Gamma$. Like y, $y' \in \Gamma$, there is $w \in \Gamma$ such that $w \leq w \leq y$, y'. However, by (A2) we have that $y, y' \leq x \wedge y' \leq x \wedge z$. Therefore, there is $w \in \Gamma$ such that $w \leq x \wedge$ z.

($\Rightarrow$). Let x, $z \in \Gamma$. Of course, $x \wedge z \in \Gamma$. Therefore, there is $y \in \Gamma$ such that $y \leq x \wedge z$.

10.2 Saturated Sets

Eytan discusses the concept of saturated sets, which correspond to the notion of realizability in logic. A saturated set is defined as a set of formulas that satisfies certain closure properties under the valuation of a logical system. Eytan demonstrates that every saturated set corresponds to an ultraprefilter, which is the maximal extension of a prefilter. This leads to algebraic completeness, where every consistent set of formulas can be extended to a realizable model.

Definition 2.1 Let $[B, \equiv, \to, \wedge, \vee, \neg, \leq, 0, 1]$ be a Boolean pre-algebra. A subset S of B is said to be saturated if the following conditions are met:

(S_1) $1 \in S$ and $0 \notin S$
(S_2) $\neg x \in S$ is and only if $x \notin S$
(S_3) $x \wedge y \in S$ is and only if $x, y \in S$.

(S_4) $x \wedge y \in S$ is and only if $x \in S$ or $y \in S$
(S_5) $x \wedge y \in S$ is and only if $x \notin S$ or $y \in S$

Definition 2.2 We call the valuation of *B* all homomorphisms of *B* in 2.

Lemma 2.3 The co-core of any valuation of *B* is a saturated set.
Proof. Let f: $B \rightarrow 2$ be a valuation. Let

$$S = f^{-1}(\{1\}) = \{x \in B \mid f(x) \equiv 1\}$$

As $f(\neg x) \equiv \neg f(x)$, it is clear that $\neg x \in S$ if and only if $x \notin S.x \notin S$.

$$f(x \wedge y) \equiv 1 \equiv f(x) \wedge f(y)$$

if and only if $f(x) \equiv f(y) \equiv 1$
Therefore, $x \wedge y \in S$ if and only if $x, y \in S$. In the same way, we get (S4) and (S5).

Corollary 2.3.1 A necessary and sufficient condition for a 2-valued function f to be a value is that the set $f^{-1}(\{1\})$ is a saturated set.
Proof. A consequence of the previous lemma.

Theorem 2.4 A saturated set of a Boolean pre-algebra is an ultrafilter.
Proof. Immediate.

Definition 2.5 Let *x* and *y* be elements of a Boolean *B* pre-algebra. We say that *x* is a logical consequence of *y* (or that from *x*, we can deduce *y*) if for every valuation f, $f(x) \equiv 1$ implies $f(y) \equiv 1$.

Lemma 2.6 x is a logical consequence of y if and only if $x \leq y$.
Proof. ($\Leftarrow$). if $x \leq y$, then $f(x) \leq f(y)$. Therefore, if $f(x) \equiv 1$, then $f(y) \equiv 1$.
($\Rightarrow$). Immediate.

Definition 2.7 An element *x* of a Boolean *B* pre-algebra is said to be a tautology of *B* if $f(x) \equiv 1$ for every valuation f of *B*.

Lemma 2.8 The set of *B* tautologies is the intersection of all saturated sets of *B*; that is, it is the class of 1 of the quotient algebra $B/\equiv$ obtained in Theorem 1.5, or else, it is the set of elements that we can deduce of 1.
Proof. Immediate.

Definition 2.9 We will say that an element $x \in B$ is satisfiable if there is at least one valuation f of *B* such that $f(x) \equiv 1$.

Lemma 2.10 The set of satisfiable elements is the union of the saturated sets of *A*.
Proof. Immediate.
Let us denote by S* (x) the family of the saturated sets of *B* that contain *x*, i.e., the family of the ultrafilters of *B* that contain *x*. Let S be the set of the saturated parts of *B*, which we canonically provide with a Boolean algebra structure.

Theorem 2.11 The application S*: $B \to$ S is a homomorphism of B in S, i.e.:
Proof. Immediate.

(H$_1$) $S^*(x \to y) \equiv S - (S^*(x) - S^*(y))$
(H$_2$) $S^*(x \wedge y) \equiv S^*(x) \cap S^*(y)$
(H$_3$) $S^*(x \vee y) \equiv S^*(x) \cup S^*(y)$
(H$_4$) $S^*(\neg x) \equiv S - S^*(x)$
(H$_5$) $\textit{if } x \leq y, \textit{ then}, S^*(x) \subseteq S^*(y)$
(H$_6$) $S^*(0) \equiv \emptyset$
(H$_7$) $S^*(1) \equiv S$

Definition 2.12 Let R be a subset of a pre-algebra of Boolean B. It is said that B is realizable (or admits a model) if each $x \in R$ is satisfiable simultaneously (i.e., by the same valuation). Likewise, it is said that from R we can deduce y, if for each $x \in R$, from x we can deduce y. We write $R \leq y$, if, for each $y \in R$, we have $y \leq x$.

When R $\leq$ 0, R is said to be inconsistent. Otherwise, R is said to be consistent.

Theorem 2.13 Let R be a part of B, the support of Boolean pre-algebra B. They are equivalent:
Proof. Immediate.

(1) R is satisfiable (model is allowed).
(2) There is a saturated set of B containing R.
(3) $S^*(R) \neq \emptyset$.
(4) There is an ultrafilter of B containing R.
(5) R is consistent.

Corollary 2.13.1 The following conditions are equivalent:

(1) B is inconsistent.
(2) $S^*(R) \neq \emptyset$.
(3) The only filter containing B is the improper filter.
(4) Every filter containing B contains 0.

Proof. A consequence of the previous theorem.

10.3 Smullyan's Tableaux

The Smullyan tableaux method, a deductive system for propositional and predicate logic, is typically seen as a procedural technique for determining the satisfiability of logical formulas. However, Eytan approaches this method from an algebraic perspective to understand how the tableaux construction can be formalized within an algebraic system. The tableaux method, particularly in its propositional form, systematically decomposes logical formulas through a tree-like structure, where formulas are broken down into simpler components until either a contradiction is found or the process halts, indicating the satisfiability of the formula.

The following rules for the construction of "tableaux" are essentially due to Smullyan:

(T$_1$) $\frac{x}{x}$
(T$_2$) $\frac{x \wedge y}{\begin{array}{c} x \\ y \end{array}}$
(T$_3$) $\frac{\neg(x \wedge y)}{\neg x \mid \neg y}$
(T$_4$) $\frac{\neg(x \vee y)}{\begin{array}{c} \neg x \\ \neg y \end{array}}$
(T$_5$) $\frac{x \vee y}{x \mid y}$
(T$_6$) $\frac{\neg(x \rightarrow y)}{\begin{array}{c} x \\ \neg y \end{array}}$
(T$_7$) $\frac{x \rightarrow y}{\neg x \mid y}$
(T$_8$) $\frac{\neg 0}{1}$
(T$_9$) $\frac{\neg 1}{0}$

We say that one of these rules applies to element x if the rule is of the form $\frac{x}{y}$ or $\frac{x}{\begin{array}{c} y \\ z \end{array}}$ or $\frac{x}{y \mid z}$.

Definition 3.1 A tableaux for an element $x \in B$ is an ordered, binary tree where the loops are elements of B and are constructed following the following processes:

(C$_1$) Place the element x at the origin.
(C$_2$) Suppose y is a peak similar to a tree at a particular stage.

(1) If one of the rules T1, T8, T9 is applicable, extend the branch where y is located and inscribe the element under the line.
(2) If one of the rules T2, T4, T1 is applicable, extend the branch where y is located and successively enter (in order) the elements under the line.
(3) If one of the rules T3, T5, T7 is applicable, branch off the branch where x is located, and inscribe on the left branch (respectively on the right) the element that is on the left (respectively on the right) of the bar, under the line.
(4) On each branch containing the elements x and $\neg x$ simultaneously, insert element 0 (in succession).

(C$_3$) On each branch that does not contain element 0, start over (C$_1$).

A branch containing element 0 is said to be closed. Otherwise, it is said to be open.

A branch is called developed if the abovementioned process is carried out as much as possible. A tableaux "for x" is called developed if each tree branch is closed or developed.

Observation 3.2 The construction of a developed branch ends after a finite number of steps if the pre-algebra in question is atomic and the degree (the number of operations affected by its arities) decreases after each rule application. Such is the case in the classical propositional calculus.

Definition 3.3 A tableaux is said to be open if it contains at least one open branch.

10.4 Hintikka Set

Hintikka sets, another key concept discussed, are named after J. Hintikka, who introduced them in his semantic analysis of quantification theory; see [2]. A Hintikka set represents a collection of formulas that satisfy certain consistency conditions, making it a vital tool in model theory for constructing models of logical theories. Eytan seeks to embed Hintikka sets within the framework of pre-Boolean algebras, offering an algebraic interpretation of how these sets function as bases for filters and ideals in logical systems.

Definition 4.1 Let $[B, \equiv, \rightarrow, \wedge, \vee, \neg, \leq, 0, 1]$ be a Boolean pre-algebra. A subset H of B is called a Hintikka set if the following conditions are met:

(H_1) $0 \notin H$
(H_2) if $x \wedge y \in H$, then $x \in H$ and $y \in H$
(H_3) if $\neg(x \wedge y) \in H$, then $\neg x \in H$ or $\neg y \in H$
(H_4) if $\neg(x \vee y) \in H$, then $\neg x \in H$ and $\neg y \in H$
(H_5) if $x \wedge y \in H$, then $x \in H$ and $y \in H$
(H_6) if $\neg\neg x \in H$, then $x \in H$

Theorem 4.2 A set of Hintikka H is the basis of an ultrafilter.

Proof. Immediate.

The Hintikka's Lemma, as applied in the tableaux method, is also reinterpreted algebraically. Eytan shows that an open branch in a tableaux corresponds to a Hintikka set, and thus, every open tableaux can be realized by a saturated set. This algebraic perspective provides a new understanding of the relationship between syntactic tableaux and their semantic models; see [2].

Corollary 4.2.1 (Hintikka's Lemma) The set of elements located on a developed and open branch is (simultaneously) satisfiable.

Proof. A consequence of the previous theorem.

Corollary 4.2.2 An x element of a pre-algebra is a tautology if and only if all tableaux developed for $\neg x$ have all the branches closed.

Proof. A consequence of the previous theorem.

Observation 4.3 The genuine interest in the results above is limited to the case of atomic pre-algebras (for example, the classical propositional calculus and C_1 Calculus);

10.5 Lindenbaum and Hintikka

Definition 5.1 Let $[B, \equiv, \rightarrow, \wedge, \vee, \neg, \leq, 0, 1]$ be a pre-algebra. A subset C of B is said to be a consistent maximal subset if C is consistent; if no part of B contains C, it is not consistent.

Theorem 5.2 Every consistent maximal subset of B is a saturated set (i.e., an ultrafilter).

Proof. We leave it as an exercise.

Corollary 5.2.1 (Lindenbaum's theorem) Every consistent subset of B can be extended to a consistent maximal set (i.e., an ultrafilter).

Proof. Immediate.

This approach highlights the importance of algebraic techniques in the study of logic, particularly for understanding the structural properties of logical systems.

As we have seen, well-established logical methods such as tableaux and Hintikka sets can benefit from a rigorous algebraic interpretation.

Note: Previous results can be extended to first-order and higher-order logics, for example, by adapting results from Rosenbloom [3] and Henkin [4].

References

1. Eytan, M.: Tableaux de Smullyan, ensembles de Hintikka et tout ça: um point de vie algèbrique. Math. Sci. Humaines **48**, 21–27 (1975)
2. Hintikka, J.: Form and content in quantification theory. Acta Philosophica Fennica **8**, 57–55 (1955)
3. Rosenbloom, P.C.: The elements of mathematical logic. Dover (2005). ISBN-10: 0486446174, ISBN-13: 978-0486446172
4. Henkin, L., Monk, J.D., Tarski, A.: Cylindric algebras, Part I, Lecture Notes in Mathematics (Book 883), p. 323 (1981). ISBN-13: 978-3540108818; ISBN-10: 3540108815
5. Gabbay, D.M.: Intuitionistic fuzzy logic and other applications. J. Philos. Log. **10**(3), 239–270 (1981)
6. Troelstra, A.S., van Dalen, D.: Constructivism in Mathematics: An Introduction. North-Holland (1988)
7. Iséki, K., Tanaka, S.: An introduction to the theory of BCK-algebras. Math. Japonica **23**, 1–26 (1978)
8. Sikorski, R.: Boolean Algebras, Springer-Verlag, 1969, ISBN-13: 978-3540044697, ISBN-10: 3540044698
9. Heyting, A.: Intuitionism: An Introduction, 2nd edn, Amsterdam: North-Holland Publishing Co. (1966)

Chapter 11
Conclusion

11.1 Summary

This monograph, An Introduction to Curry Systems, sought to formulate a unified framework for the algebraization of non-classical logics (among others), with special emphasis on paraconsistent, paracomplete, and non-alethic systems [1]. By developing the notion of Curry Systems, we provide a conceptual and formal basis that allows the algebraic modeling of logics that reject or relativize classical principles, such as the Principle of Non-Contradiction and the Principle of Excluded Middle [2, 3].

At the heart of this investigation is the recognition that the algebraic representation of logic does not need to be restricted by the rigid conditions of classical equivalence of equality. Instead of requiring formulas to be grouped into equivalence classes modulo provable equivalence (as in the traditional Lindenbaum-Tarski approach), we construct algebraic systems based on prestructures. These include pre-lattices, pre-Boolean algebras, pre-Heyting algebras, and several other pre-algebraic systems sensitive to syntactic distinctions between formulas, even when they are logically equivalent. This methodological choice allows us to directly encode intensional and inferential aspects of formulas without compressing them into identical algebraic elements.

Throughout the text, we introduce and systematically develop the algebraic counterparts of several important logical calculi. In particular:

C_n-algebras: These structures algebraize da Costa's paraconsistent C_n-calculus, capturing the essential feature that the negation operator is incompatible with the fundamental equivalence relation. In this context, negation behaves as a non-monotonic operator that is not necessarily involutive or classically contradictory. C_n-algebras illustrate how contradictions can be tolerated algebraically without resulting in trivialization.

P_n-algebras: Analogously, P_n-algebras provide a formal structure for paracomplete logics, where the Principle of Excluded Middle fails. Here, the logic allows

J. M. Abe, *An Introduction to Curry Systems*, Intelligent Systems Reference Library 280, https://doi.org/10.1007/978-3-031-97810-4_11

for the possibility that neither a proposition nor its negation is provable, and the corresponding algebraic structures reflect this indeterminacy through non-classical lattice operations and weakened completeness conditions.

N_n-algebras: These algebras possess features of both paraconsistent and paracomplete systems, interpreting connectives completely flexibly. N_n-systems are, therefore, well suited to modeling reasoning under uncertainty, vagueness, or incomplete information, where truth values are context-sensitive and logic becomes non-referential.

C_n^*, P_n^*, and N_n^*-algebras: By extending propositional Curry systems to first-order logic, these algebras include operations for quantification (existential and universal) and binding of variables. We discussed how quantifier semantics should be redefined in the absence of classical assumptions and how algebraic semantics can be maintained even in non-classical first-order contexts [4].

$Q\tau$-algebras: These structures generalize the annotated systems and provide a unifying framework for embedding classical and non-classical systems. Such systems are, in general, non-alethic.

In addition to presenting these algebraic systems, the book also explores the duality and interaction between logical syntax and semantic algebraic structures. This exemplifies the role of congruence relations, homomorphisms, and the preservation of logical consequences in algebraic applications [3]. We showed how the completeness, correctness, and compactness theorems can be reformulated within Curry's algebraic framework, thus extending the reach of traditional model theory.

One of this work's main conceptual contributions is that algebraic semantics need not rely on equivalence classes or quotient constructions to capture logical consequences. Instead, we can use algebraic pre-structures that retain more information about the derivational history or syntactic form of formulas. This allows for finer-grained modeling of logics in which syntactic distinctions are semantically relevant, such as in paraconsistent or epistemic contexts.

Another important insight is the role of non-involutive negation in Curry Systems. While classical negation is defined as an involutive operator that satisfies $\neg\neg p \equiv p$, negation operators in Curry Systems are context-dependent and do not need to satisfy such identities. This allows for modeling various modal, epistemic, and substructural logics requiring non-standard negation forms.

The book also highlighted the philosophical and computational importance of Curry Systems. In logic programming, knowledge representation, and formal ontology, systems that tolerate inconsistency or incompleteness are essential for dealing with information from the real world.

The algebraic perspective presented here provides tools for reasoning in such systems with mathematical rigor.

It is worth noting that this work represents a first fundamental approach to Curry Systems. Much remains to be explored. In future research, we foresee the development of categorical models of Curry Systems, possibly based on enriched categories, indexed categories, or co-algebraic structures. We also foresee applications (as an algebraic system) in areas such as artificial intelligence, belief revision, information fusion, and legal reasoning, where non-classical logics naturally arise.

In addition, the treatment of quantifiers, modalities, and intensional operators in Curry Systems can be extended to richer languages, including higher-order logics, dynamic logics, and hybrid systems. Using topological and algebraic geometry tools to analyze non-classical structures can yield new insights into logical continuity, connectivity, and space.

In conclusion, this book has provided a theoretical framework that redefines the algebraic foundations of non-classical logic through Curry Systems. We have laid the foundation for more nuanced and expressive algebraic semantics by abandoning the strictly equivalence-centered approach of classical algebraic logic. The results and insights presented here are theoretical artifacts and a platform for future research and innovation in logic, mathematics, and computer science.

11.2 Future Directions

Building on the foundations presented in "An Introduction to Curry Systems", several promising avenues exist for future research. One of the most important directions is the categorical reformulation of Curry Systems. Category theory, particularly its enriched and indexed variants, offers powerful abstractions that can help unify the diverse algebraic structures presented in this work. Categorical semantics can provide a deeper understanding of morphisms between Curry algebras, natural transformations between semantic functors, and adjoint situations involving logical consequences and structural operators.

Another direction concerns the expansion of Curry Systems to co-algebraic structures, especially for modeling dynamic and nondeterministic processes. Since Curry Systems are particularly well suited to modeling paracomplete, adversarial, and context-sensitive reasoning, they can serve as semantic foundations for belief revision systems, information flow, and dynamic epistemic logic. Coalgebras can provide a complementary perspective by emphasizing the evolution of states and transition systems, thus extending the applicability of Curry algebraic tools to areas such as distributed systems and temporal reasoning.

Another important work is the development of Curry Systems for modal and intensional logics, where traditional equivalence-based semantics often fail to preserve critical distinctions for modal reasoning [5]. Here, the emphasis on syntactic form and inferential structure becomes a virtue, allowing the definition of fine-grained accessibility relations and intentional operators. Modal Curry algebras and intensional extensions would offer a rich field of exploration.

In terms of practical applications, Artificial Intelligence and knowledge representation remain fertile ground for deploying Curry Systems. In contexts where data may be inconsistent or partial, such as legal reasoning, ethical decision-making, and multi-agent systems, Curry algebraic semantics provide a coherent formalization means. In particular, annotated evidential reasoning, paraconsistent information fusion, and paracomplete inference mechanisms could benefit from a Curry-style algebraic foundation.

Another area ripe for exploration is using topological and sheaf-theoretic tools in modeling Curry Systems. Stone duality and sheaf semantics have long established the connection between topology and logic [6]. Curry systems could benefit from topological representations where truth values vary across contexts or locations and logical operators correspond to continuous transformations in structured spaces [7]. Such an approach could open up new perspectives in spatial logic and context-sensitive reasoning.

Finally, computational implementations of Curry systems will be crucial to test their effectiveness and extend their reach. Building proof assistants, automated theorem provers, or reasoning engines based on Curry's algebraic principles could help validate their formal robustness and discover new inferential patterns. Integration with existing logic frameworks such as Coq, Agda, or Isabelle/HOL could accelerate the development of a Curry algebraic software ecosystem.

In summary, the potential for scaling Curry Systems is substantial theoretically and in application. This book has provided a conceptual starting point; what remains is the collaborative effort to refine, extend, and apply these ideas across disciplines. The path forward promises profound mathematical insights and impactful contributions to logic, computation, and philosophy.

References

1. Abe, J.M.: An introduction to the theory of Curry Systems, Scientiae Mathematicae Japonicae, *Scientiae Mathematicae Japonicae,* **76**(2), 175–194: e-2013, ISSN 1346–0862 (2013)
2. Abe, J.M., Nakamatsu, K.: Curry systems for algebraization of some non-classical logics. Int. J. Reason. Based Intell. Syst. ISSN: 17550556, Interscience Enterprise, UK, 59–69 (2011)
3. Rasiowa, H., Sikorski, R.: The Mathematics of Metamathematics. Polish Scientific Publishers (1970)
4. Abe, J.M., Nakamatsu, K., Akama, S.: An algebraic version of the monadic system C_1, new advances in intelligent decision technologies. Series: Studies in Computational Intelligence, vol. 199. In: Nakamatsu, K., Phillips-Wren, G., Jain, L.C., Howlett, R.J. (eds.), 16, pp.641–349, Hardcover, Springer Berlin, Heidelberg, ISBN: 978-3-642-00908-2, (2009). https://doi.org/10.1007/978-3-642-00909-9
5. Dummett, M.: Elements of Intuitionism. Clarendon Press (1977)
6. Goldblatt, R.: Topoi: The categorial analysis of logic. North-Holland (1984)
7. Johnstone, P.: Stone Spaces. Cambridge University Press (1982)